Fire Snake

By the Same Author

Africa's Flamingo Lake
The Freshwater Eel
Nature's Light,
The Story of Bioluminescence
The Red Sea
The Sargasso Sea,
An Ocean Desert
Sea Turtles
A Secret Language of Animals,
Communication by Pheromones
Sounds in the Sea

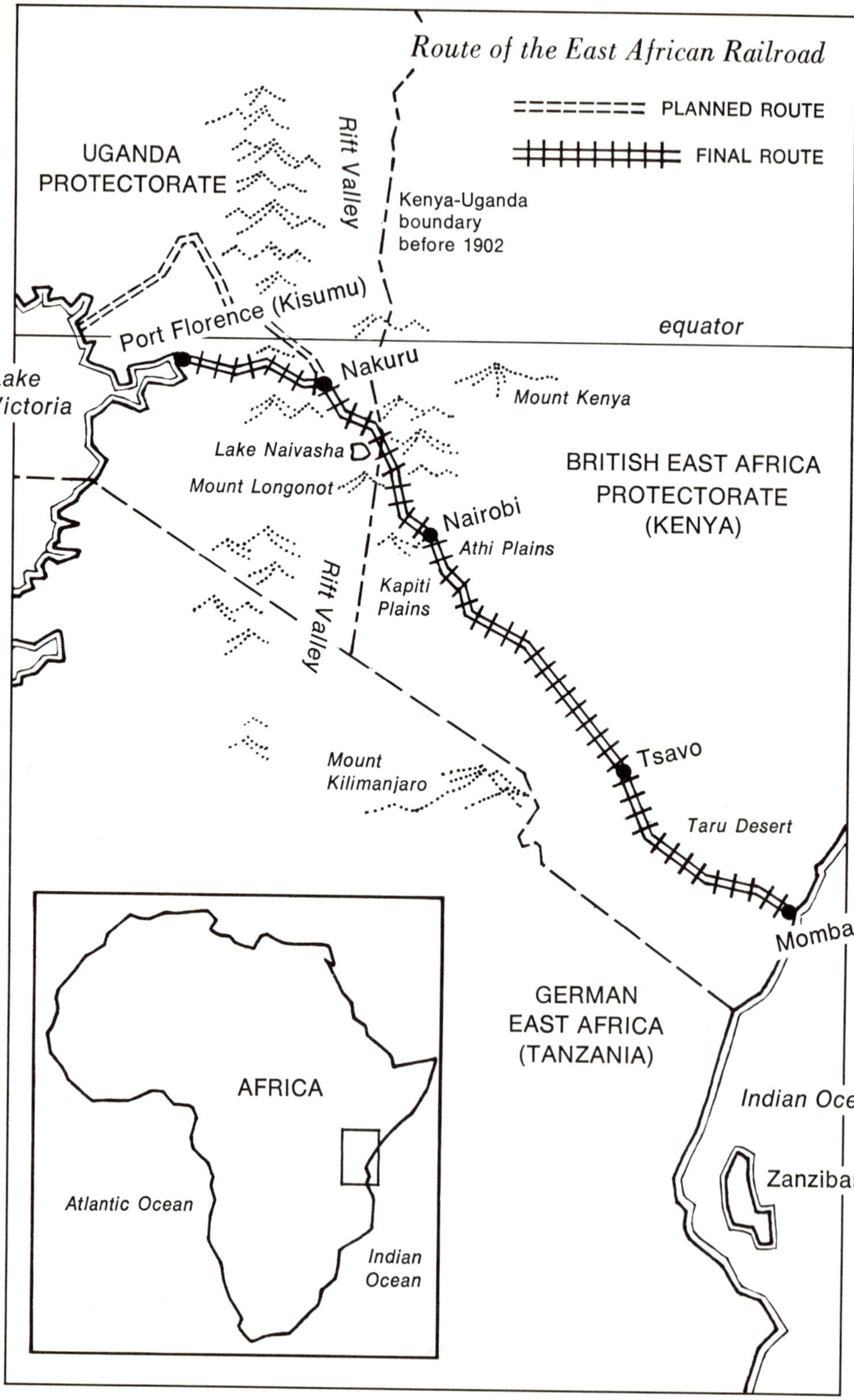
Route of the East African Railroad
PLANNED ROUTE
FINAL ROUTE
UGANDA PROTECTORATE
Rift Valley
Kenya-Uganda boundary before 1902
Port Florence (Kisumu)
equator
Lake Victoria
Nakuru
Mount Kenya
Lake Naivasha
Mount Longonot
BRITISH EAST AFRICA PROTECTORATE (KENYA)
Nairobi
Athi Plains
Kapiti Plains
Rift Valley
Tsavo
Mount Kilimanjaro
Taru Desert
Mombasa
GERMAN EAST AFRICA (TANZANIA)
AFRICA
Atlantic Ocean
Indian Ocean
Indian Ocean
Zanzibar

FRANCINE JACOBS

Fire Snake

The Railroad That Changed East Africa

with illustrations by Lloyd Bloom,
photographs, and maps

William Morrow and Company
New York 1980

Printed in the United States of America.
1 2 3 4 5 6 7 8 9 10

Library of Congress Cataloging in Publication Data

Jacobs, Francine.
Fire snake.

Includes index.
Summary: Describes the construction of the East African railroad at the turn of the century and the political and social effects of it on colonial Africa.
1. Railroads—Kenya—Juvenile literature. [1. Railroads—Kenya.
2. Kenya] I. Bloom, Lloyd. II. Title.
TF119.K4J32 625.1'009676'2 80-13840
ISBN 0-688-22232-3
ISBN 0-688-32232-8 (lib. bdg.)

The author wishes to thank Dr. John Spencer, Dean of the College, Middlebury College, for reading and checking the manuscript of this book. This author is also grateful to the Kenya Railways Corporation for the photographs and to Brian T. Yonge for his help in collecting them.

To Michael Fritz

Contents

Fire Snake

An Ancient Prophecy

The sun rose above the Indian Ocean, its rays brightening the sky over East Africa. In the fertile, green highlands of the Kikuyu tribespeople, a great medicine man named Mogo awoke, trembling after a frightening dream.

His family summoned the elders of the tribe. "Gather around and listen carefully," the old wise man urged. The elders moved closer, eager to hear the medicine man who had special powers and could predict the future. "I have had a terrible vision," Mogo began. "Last night I dreamed that Ngai (God) came and led me to a place I did not know. There he revealed to me what lies in store for our people." The elders murmured worriedly. Mogo continued. "A famine will come to this land, and afterward strange people will appear. They will come from the big water in the east, and they will have skin as white as the bellies of frogs and will dress in colors of the wings of butterflies.

"The strangers will carry magic sticks that spout fire. They will bring a great iron snake with as many legs as a centipede. It too will spit fire. The iron monster will stretch a great distance from the big water in the east to another big water in the west. The Kikuyu must not fight these strangers," Mogo warned. "Their magic sticks will kill faster than poisoned arrows. But do not welcome them either, for the white-skinned strangers will take our land away."

From an account by Jomo Kenyatta
in *Facing Mount Kenya*

1

The Iron Snake

Until the middle of the nineteenth century, the great continent of Africa was still largely unexplored by Europeans. Its vast interior was a mystery to most of the world. Only Swahili and Arab dealers, who had traded for ages along the Indian Ocean coast of East Africa, had penetrated deeper inland. The Swahilis, a mixed race descended from Arabs and coastal Africans, and the Arabs came in search of ivory and slaves.

The part of East Africa that today is the nation of Kenya was a sparsely populated near wilderness at that time. Some forty different tribes, including the Kikuyu, the Maasai, the Kamba, and the Nandi eked out a simple existence. Some hunted in the game-rich forests, some farmed, while others wandered, grazing their herds in the bush and grasslands. They lived according to the old, proven ways of their peoples, using handmade tools and weapons. They were unaware of the inventions and machines that were changing Europe and America. In interior East Africa there were no towns, no roads, no railways. The people had no carts or wagons, nor did they use horses or other animals for transport. All had to be carried on their heads or backs. Those who heard of the medicine man's prophecy about the iron snake could not have imagined that it might mean a railway.

Life was unhurried in this land without clocks or calendars. No written language existed, and it was often difficult, if not impossible, for members of one tribe to communicate with another. Hence, there was little trade except for barter. Also, there was no concept of money. People worked for themselves or helped in tribal efforts, but no one worked for wages. Tribes and clans within tribes raided each other for cattle and slaves or to settle disputes. Traditions based on spiritual beliefs, magic, and ancestor worship guided tribal life.

Arab and Swahili traders from the nearby island of Zanzibar had established a profitable commerce in ivory and human beings from the interior of East Africa even before the nineteenth cen-

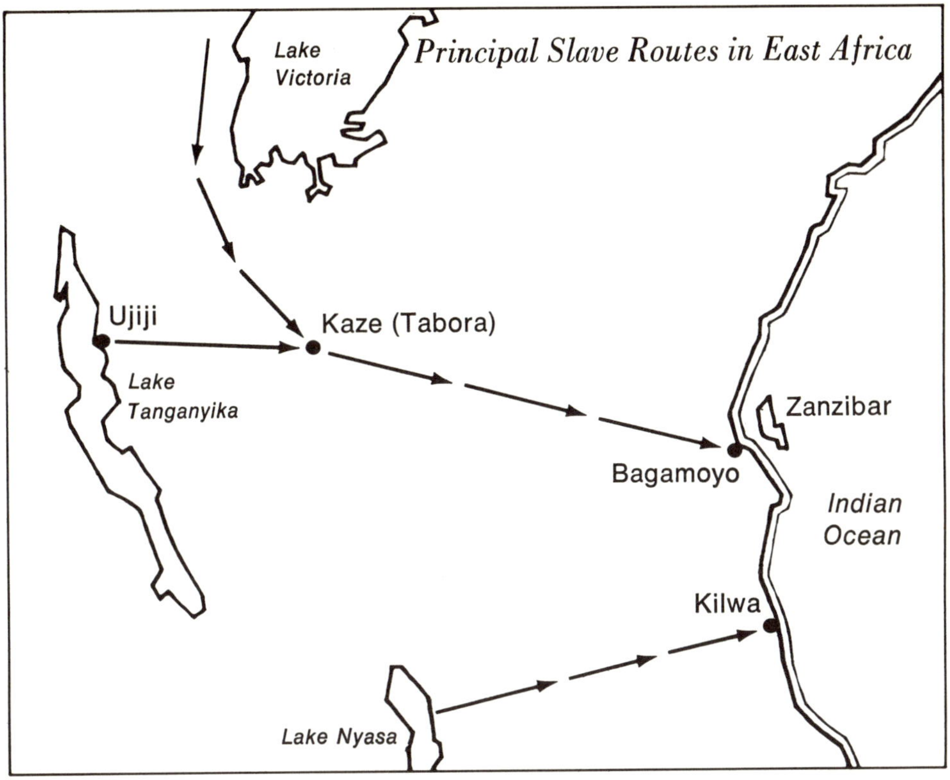

tury. Over the years, the demand for ivory, used for jewelry, ornaments, furniture, and piano keys, grew in Europe and Asia. At the same time, there continued to be a market for East African slaves in Arabia and Persia, though slavery had become illegal in the British Empire in 1833, and the traffic in slaves from West Africa to Europe and America was dying out.

The profits to be made from both ivory and slaves encouraged dealers to make still deeper forays into central Africa. They penetrated hundreds of miles; a few reached the southern and western shores of Lake Victoria. The traders bribed tribal chiefs with guns, beads, or calico cloth to raid villages for slaves. Fifty pieces of cloth might be traded for a man and as much as two hundred for a young girl. Whole villages were taken. Once captured, the slaves were moved to the sea along caravan routes through present-day

Tanzania and Mozambique. By the mid-nineteenth century, the slave traders had found a shorter route to Uganda through the territory that forms the southwestern corner of modern Kenya. This land belonged to the Maasai, nomadic warrior-herdsmen.

Lacking roads and vehicles, the traders marched their slaves along trails. Three or four traders would oversee a caravan consisting of several hundred to a thousand slaves. Chained to one another, the captives were forced to walk single file for hundreds of miles, carrying supplies and precious ivory tusks on their heads and shoulders. An ivory tusk might weigh one hundred pounds.

The slaves trudged for months under a scorching sun without proper food or enough water. Disease and the physical ordeal took a heavy toll. Vultures and hyenas gathered to feast on the bodies of the dead in the wake of the caravans. Death was so common that only one in five slaves seized in the interior reached the coast. Those who did were crowded into waiting ships. Conditions on these wooden sailing vessels, known as dhows, were horrible. The slaves were jammed into narrow racks below decks. Many who had survived the trek to the coast perished in these terrible slave ships. The remaining ones were auctioned off at the slave market in Zanzibar to work the clove plantations. Though the sultan of Zanzibar made the export of slaves unlawful in 1845, the trade went on anyway. Slaves were hidden in caves in thick underbrush and were shipped out at night in dhows to buyers in Arabia, the Persian Gulf states, and Egypt.

While slavers continued to plod the caravan trails, European missionaries and explorers were probing Africa. They came to spread Christianity and to make discoveries. The famous missionary-explorer, Doctor David Livingstone, came from Scotland to aid the sick and to find the source of the Nile River. The riddle of the Nile brought other explorers, too, including the English adventurers Richard Burton and John Speke. Joseph Thomson, a Scot, braved the dangers of crossing Maasai land to reach Lake

Victoria. These and other courageous men walked the same caravan trails the slavers used. What they saw appalled them, and their reports about the brutal slave trade were widely publicized in their homelands and elsewhere.

The horrors of the East African slave trade jolted the conscience of the Western world, and public opinion demanded that it end. The British Government sent frigates into the coastal waters of East Africa to intercept slave ships and free their captives, but the trade was so widespread that these patrols had little effect. Finally, a series of treaties between Great Britain and the sultans of Zanzibar succeeded in closing the merciless slave market there. Some slave dealers managed to continue smuggling their human cargoes out of East Africa, however.

The slave trade was actually declining, but additional measures seemed necessary to stop it. In 1892, a railway from the coast of East Africa to Uganda was proposed in the British Parliament as the most effective way to curtail the commerce in slaves further. A railroad that could carry great loads of ivory and hides quickly and inexpensively would put an end to the plodding, inefficient slave caravans.

So it was that thousands of miles away from East Africa, white-skinned strangers were discussing a railway—an iron snake—that would stretch from the sea in the east across the lands of the Kamba, the Kikuyu, the Maasai, and the Nandi to Lake Victoria in Uganda.

An old medicine man's prophecy was coming true.

2

The Lunatic Line

Was it practical for Great Britain to build a railroad in such a sparsely populated place? In 1892, members of the House of Commons began to debate this question. The leader of the Government, Prime Minister Lord Salisbury, and most of the men in the House of Commons favored the proposed railroad to Uganda as a great humanitarian step. They hailed it as a blow to the evil slave trade and also as a measure that would benefit the people of East Africa by bringing modern ways to them.

Opponents of the railway did not share this enthusiasm; they worried about the cost and who would pay for it. The opposition was led by Henry Labouchere, an able debater respected in the House of Commons for his intelligence, wit, and boundless energy. Labouchere fought the railway vigorously, denouncing it as a noble-sounding but foolish venture. He was quick to point out that despite their eloquent words and sentiments the railway's supporters understood little about the people of East Africa or their needs. He accused those who favored the railway of having another motive.

Labouchere believed that the railway was a convenient way to extend the British Empire farther into Africa. He and his followers opposed expansion of the Empire. They proclaimed that Britain had no right to build this railway on the soil of another country. Britain, they said, should stay out of Africa.

The debate was heated; tempers were aroused. Henry Labouchere's biting speeches did little to calm the atmosphere in the House of Commons. He denounced the Government's case as so much "vague, general claptrap." He also attacked those who disregarded the rights of Africans. "They may be bloodthirsty, they may be savages, or what you like; but you have no right to go into a country and drive off the inhabitants because you are pleased to call them bloodthirsty."

The debate over the railroad dragged on for several years. Opponents fought it in Parliament and in the press, mainly with the weapon of ridicule. The proposed railroad was to be called the Uganda Railway, but its enemies named it the Lunatic Line.

While the debate continued, a team of surveyors was sent to East Africa. They returned in 1893 to report that a single-track line from Mombasa on the Indian Ocean coast of East Africa to Lake Victoria in Uganda could be built in four years or less at a cost of three million pounds sterling.

The report was welcomed in Great Britain. Many Englishmen looked forward to opening East Africa to trade. The British had established an economic foothold in Uganda; a private company operating under a Royal charter was already conducting business there. When this enterprise failed, Britain officially planted her flag in the territory and declared Uganda a protectorate in 1894. A protectorate is a weak land under the protection, or control, of a stronger one.

Uganda was fertile and believed to be rich in resources. A railroad could bring its wealth down to the coast for shipment to the world's markets. The British worried that if they delayed, Germany, which controlled the territory south of Uganda (Tanzania today), would build a railway there first. In the 1890s, many Englishmen felt that their nation had to compete successfully with Germany and with other European powers, such as Belgium, France, Portugal, and Italy, which were busily acquiring possessions in Africa. A railway to Uganda, which would carry British troops quickly to the interior and to the headwaters of the Nile River, would strengthen British influence in the Sudan and Egypt. Most important of all, it would help make the Suez Canal, Britain's lifeline to India, more secure.

The British Government wanted to control the land through which the railroad would pass. This territory, which lacked resources and was largely underdeveloped, had been administered by the same British chartered company that failed in Uganda. In 1895, Britain took this land as well, declaring it the East Africa Protectorate. The area came under the control of the British Foreign Office, which governed it from London.

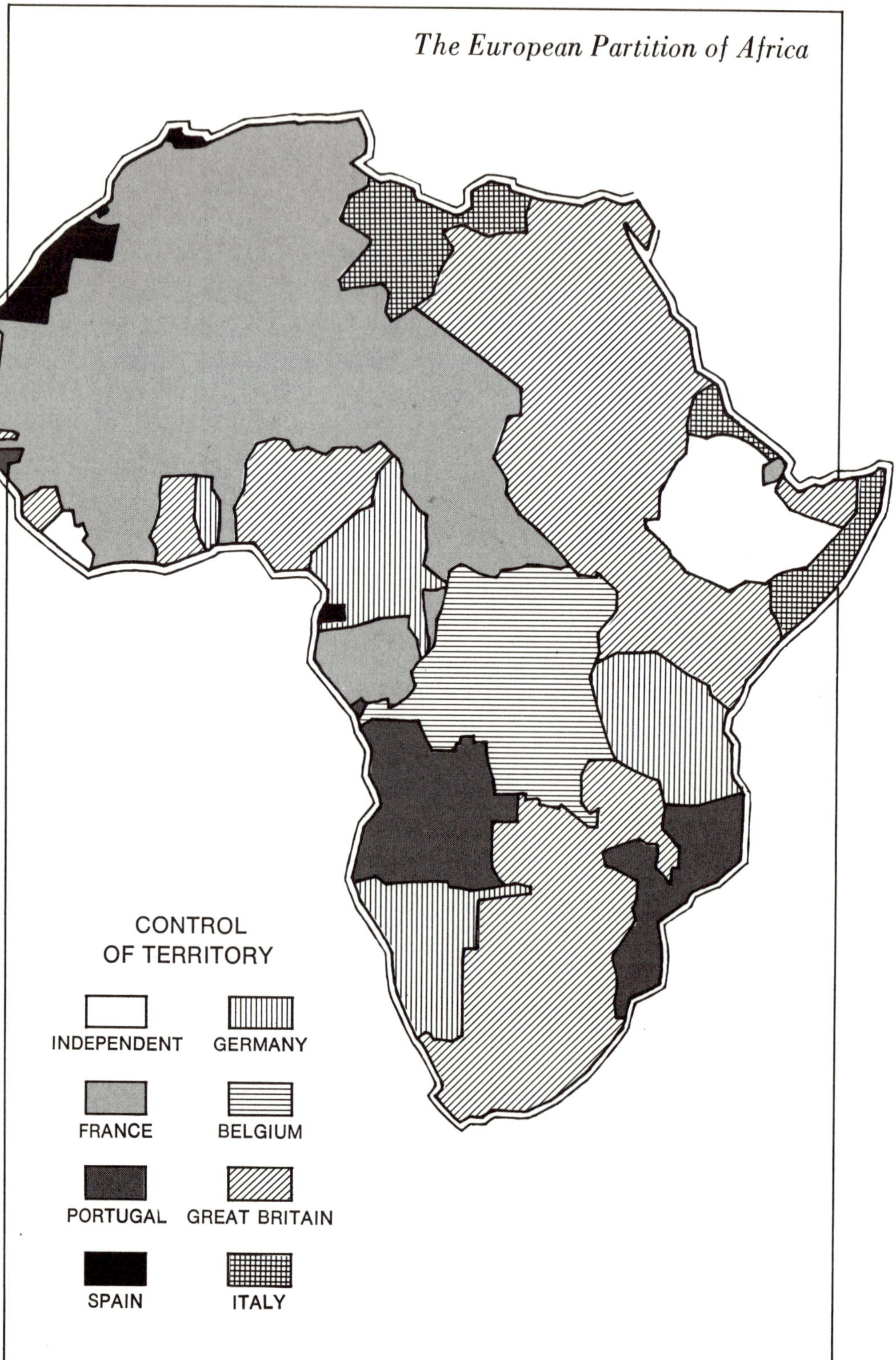
The European Partition of Africa
CONTROL
OF TERRITORY
INDEPENDENT
GERMANY
FRANCE
BELGIUM
PORTUGAL
GREAT BRITAIN
SPAIN
ITALY

By 1895, it was clear that Parliament would finally approve the proposal for a railroad in East Africa, and so the British Government appointed a Railway Committee of Foreign Office officials to oversee construction. The Committee, in turn, hired a chief engineer to direct the building. The Englishman chosen, George Whitehouse, had built railroads in England, Mexico, South America, South Africa, and India.

On December 11, 1895, the steamship *Ethiopia* brought George Whitehouse into the harbor of Mombasa. He was a tall, neatly dressed man in his mid thirties, with the confident manner of an upper-class British gentleman. Weeks earlier, Whitehouse had left cold, gray England; now under a brilliant sky his ship passed beneath the rusting cannons of Fort Jesus, a stone fortress built by the Portuguese three hundred years earlier.

The narrow, shallow harbor of Mombasa was crowded with dhows. These wooden sailing vessels and their Arab masters came on the *kazkazi,* the seasonal monsoon winds that blow steadily across the Indian Ocean toward the African coast between October and April. For centuries dhows had arrived laden with carpets, silverware, brass, tiles, dates, and spices at this time of year. Months later they set sail again on the *kuzi,* the tradewinds that blow eastward from April to October, carrying ivory, rhinoceros horn, skins, coconut oil, timber, and slaves to Arabia and the Persian Gulf.

The ancient island city of Mombasa, separated from the mainland by a creek, shone in the tropical sun. Coconut palms and thick-trunked baobab trees framed flat-roofed, white stucco houses. Coconut plantations covered much of the island in 1895. The old part of the city began at the wharves and reached uphill into a maze of narrow, twisting alleyways, streets, and tall houses. Arabs lived in this quarter. Other sections of the city sheltered Africans, Swahilis, and a few Englishmen, though Arabs and Swahilis made up the bulk of Mombasa's population.

Whitehouse boarded a tiny trolley, a *gharri,* which ran from the shallow old port on the eastern side of the island to Kilindini, the deepwater harbor at its western end. The tramway had been installed by the British chartered company that had administered the territory before it had become part of the British Empire. Two strong Swahilis pushed the single, canopied car, which had room for four passengers seated back to back on a narrow bench. Many English gentlemen might have found the stench of the open sewage system sufficient reason to leave Mombasa as soon as possible, but Whitehouse was undeterred and settled down to his task. His lodgings were at the Grand Hotel, old and shabby, despite its name.

The earlier railway survey team had supplied Whitehouse with inadequate maps and sketches, which lacked important information about the terrain, details about curves, grades, and elevations. Although another survey was supposed to have taken place, it had not been done. The first rough charts showed the railroad's route generally following a caravan track—the shortest path to Uganda.

The proposed railway was to be 657 miles long and run northwestward from Mombasa inland to Lake Victoria. It would cross desert, dense bushland, mountainous terrain, and Africa's Great Rift Valley, an enormously wide split in the surface of the earth. It would pass through grassy plains and boggy marshland before reaching its destination.

Whitehouse recognized that the foot trail the railroad was to follow would not always be suited to trains. Unquestionably, the railway would have to depart from the caravan route here and there and run through land the surveyors had not actually explored. The missing information as well as his own firsthand knowledge of the problems involved in constructing railroads made Whitehouse doubt the surveyors' conclusion that there were "no great or serious difficulties to overcome" in building the line. The Great Rift Valley with its steep walls alone would be an enormous obstacle. Clearly the construction of the Uganda Railway promised to be an adventure.

3

Nature Takes a Toll

In the months since George Whitehouse had established himself at the Grand Hotel, he had become accustomed to its dirt, ants, broken bamboo furniture, and noises—the clacking of beaded curtains and the sounds that came into the poorly lit lounge from the billiard room and bar. He used these months to set up a base of operations in Mombasa.

Whitehouse concentrated on organizing a staff and preparing a depot to handle the railway's traffic and supplies. Nearly everything that would be needed—steel rails, crossties, metal plates and bolts for fastening, spikes and hammers—would be brought in by ship. These cargoes would have to be off-loaded and stored in an orderly way so that each item would be available when required. The old, shallow, dhow-crowded harbor could not receive the large steamships that would bring in the masses of supplies and equipment or the locomotives and cars that would roll on the railroad, so Whitehouse purchased land for the depot near the deepwater harbor at Kilindini. Under his direction offices, warehouses, workshops, tool sheds, housing for workers, even a hospital were built on some 372 acres of coconut plantations. Construction of a wooden jetty with tracks for a mobile crane began. It was needed to help off-load the heavy cargoes that had started to arrive.

Though the depot was ready in four months, Whitehouse still worried about time. Thousands of miles away, Parliament, at first reluctant to build the railway, now was impatient to see it under way. It wanted the construction of the estimated 657-mile railroad to be completed in less than four years.

Parliament did not understand the conditions in East Africa that Whitehouse faced, however. Labor was a problem. The population in the area where the railroad would be built was sparse, and local Africans were unfamiliar with picks, spades, and the tools of railroad construction. Moreover, they were unaccustomed to working for wages, and most were hesitant or unwilling to leave their farms and herds. The few Africans who agreed to work on the railroad at first were paid in goods—wire, knives, and food. They were used as water carriers or porters and hauled supplies on their heads to the construction gangs.

Most of the laborers for the railway came from India. The British had built railroads in India, and many workers there were skilled. Now the promise of jobs on the Uganda Railway attracted Indian draftsmen, stonemasons, carpenters, surveyors, and clerks. Thousands of unskilled laborers were recruited also.

Water was another problem. It was needed to quench the thirst of workers laboring under the hot, equatorial sun and also to make the steam that drove the construction tractors and locomotives. Water was precious in East Africa, where there are long dry seasons between rains. As it was, the residents of Mombasa had barely enough for themselves. The railway would have to find its own source. A stream was dammed and a reservoir constructed to hold water that fell in the rainy seasons. Later Whitehouse set up an installation to distill fresh water from the sea. He planned to send daily water trains from Mombasa to crews laying the first hundred miles of track northwestward to a place called Voi. There a river would provide water. But first he had to reach the mainland.

The deep, narrow channel that separated the island of Mombasa from the mainland of East Africa was called Makupa Creek. Whitehouse at first had planned to get over the creek by creating an earthen embankment across it. He discovered, however, that dhows and other small ships used the waterway, and he could not block the channel and deny them passage. He would have to build a bridge to get his railway to the mainland. Two bridges, in fact. The first, a wooden one, would serve temporarily until a permanent steel bridge could be put up.

Whitehouse quickly drew up plans for the wooden bridge. When he set out to obtain the timber he needed, however, he discovered no one could supply it. Finally he learned that trees of a type he could use were to be found a few hundred miles inland. But there was no access to them. Frustrated, Whitehouse sent off a cable to London. He would have to import his lumber all the way from England. Narrow, unimpressive Makupa Creek, the first obstacle in the path of the railway, was teaching Whitehouse an African lesson—patience.

It took four months, from January to April, 1896, for the wood to arrive. By then the longer of the two rainy seasons had begun. As fate would have it, these rains, which usually end in June, continued for months, slowing construction of the bridge. What should have taken three weeks to complete took the workers almost four more months, until August. Nature had exacted a toll in precious time.

On the mainland at last, the Uganda Railway was set to advance. The first engines and railroad cars arrived from England and were unloaded at the depot. Whitehouse chose a locomotive, a tender, and two cars to be his personal train. One car would contain an office and the other his living quarters as he traveled back and forth from Mombasa to the railhead, the construction site. He had his train painted bright yellow, red, and green. On the tender, which carried coal and water behind the engine, the letters U.R., for Uganda Railway, were painted in gleaming gold. Notwithstanding all the years he had worked on railroads, George Whitehouse could not conceal his excitement with his shiny, new train.

The railway paid nothing for the land it used on the mainland, and the cost of construction materials was more or less as projected. Time, however, proved uncontrollable and became the determining factor in the amount spent on the line. Any significant

---

Indian railway workers, men atop a steam-driven tractor, and one of the spectacular early accidents.

1204

delay increased the costs of construction because crews had to be paid whether working or not until the railroad was finished. The expense of the project in East Africa was already the subject of further wrangling and debate in Parliament. The fiery critic, Labouchere, had taken the floor to charge that the railway that "started from nowhere and . . . went nowhere" would end up costing five million pounds rather than the three million pounds originally forecast. Despite opposition by a determined few, the bill allocating funds for the railway was passed in August, 1896, by an overwhelming majority; the vote was 255 to 75. Parliament wanted the first hundred miles of track to Voi completed by March of the following year, 1897. It did not seem an unreasonable expectation.

The terrain rises steadily between the coast and Voi. As soon as the construction gangs arrived on the mainland, they were faced with a series of hills, rising 560 feet in the first fifteen miles. These hills were thick with tough, unyielding, thorny bush, which had to be painstakingly removed by hand. Africans wielding *pangas*, broad, double-edged knives, helped in this tedious work. Progress was slow; only about a quarter of a mile could be cleared each day. Behind the ground-clearing crews came other workers to move earth and rocks and to build embankments to receive the tracks.

Each morning at dawn supply trains from the depot in Mombasa brought materials to the railhead. From sunrise till sunset, an assembly line of workers labored. In relays, teams laid the railroad crossties on the embankments. Other workers placed the steel rails on the ties. More laborers aligned the rails and hammered them down with steel spikes. Slowly the railroad moved forward: crossties, rails, fastening; crossties, rails, fastening; crossties, rails, fastening.

A telegraph line was put up along the route. The laborers lived in temporary camps; as the work advanced, their tents, light beds, mattresses, blankets, cooking pots, meager possessions, and tools were shifted ahead.

After three months of arduous work in the rain, construction had only reached Mile 13. There a 400-foot-wide gully had to be spanned. Timbers were brought in to build a bridge. Despite the rain, the workers managed to complete the bridge in twenty-five days, but nature, which seemed to oppose the railway from the very start, appeared more determined than ever now to resist. The rains intensified; torrents fell. Only thirty-four inches of rain had fallen in the area all the year before (1895), but now, in November alone, twenty-seven and a half inches came down. The deluge made further progress impossible.

Whitehouse arrived on his personal train to inspect conditions at the railhead. He slogged his way through the downpour and thick mud to examine washed-out embankments; he spoke to foremen and looked at the wet, discouraged work crews. Back aboard his train, he drank hot tea and listened to the steady torrents drilling on the roof. A strong hatred for the rain grew within him. He realized that progress was impossible under these conditions. Reluctantly Whitehouse ordered the tracklaying stopped. Though it went against his nature to give in, the weather forced him to wait three precious weeks. By year's end a record sixty-six inches of rain had fallen, and the railhead was only at Mile 23. At this rate, the railway would take almost twenty-nine years to complete.

4

A Dread Unearthly Place

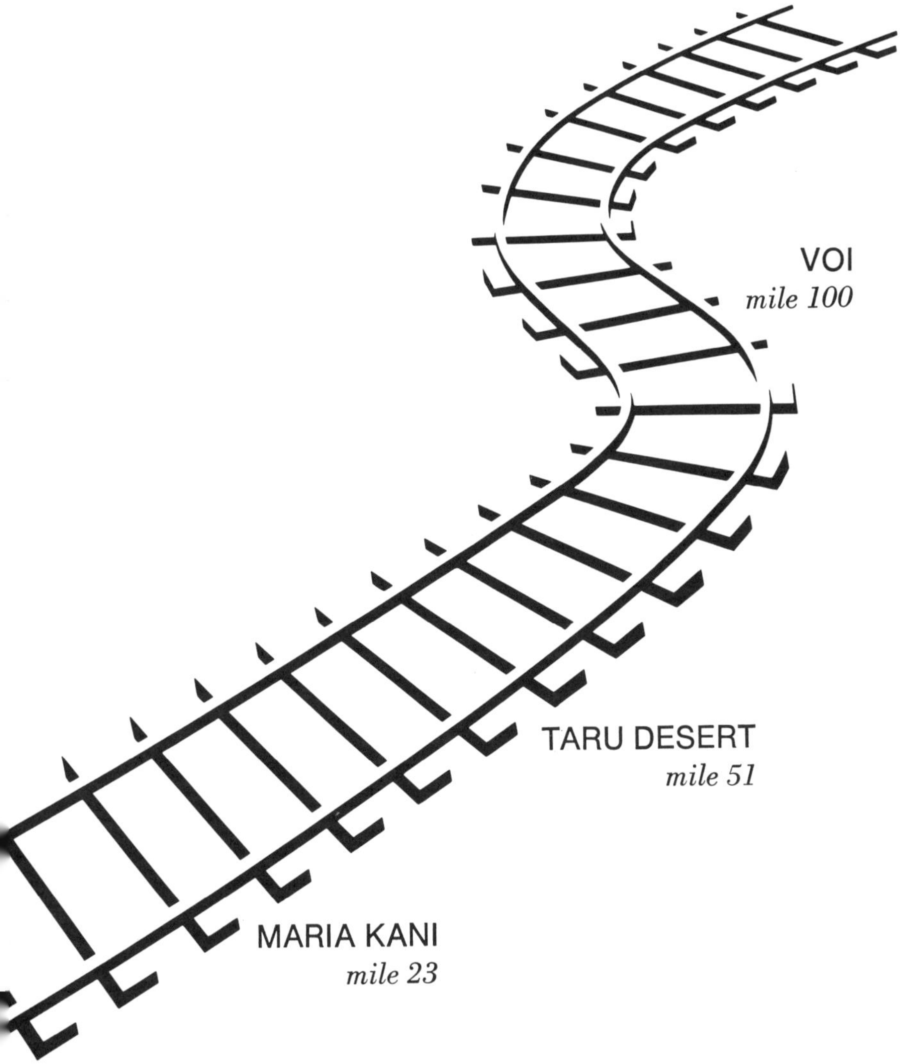

The rains continued mercilessly day after day, not only slowing construction, but creating pools of water in which malaria mosquitoes bred. At the start of 1897, half the 4000-man labor force was sick with malaria and unable to work. Men wracked with fever, breathing heavily, and shaking with malarial chills lay on beds in crowded hospital tents. Others nursed wounds and painful sores from thornbushes and jagged tree stumps. Some had ugly, bloated, festering skin ulcers, where tiny insects called "jiggers" had injected their eggs. The workers' tents stank with the excrement of those too weak from dysentery and diarrhea to help themselves.

Morale in the railway camp was low. Indian laborers shivered in their tents, playing cards or listening to the sound of the rain beating down. Unable to dry their soggy *dhotis*, the thin, white, cotton trousers and blouses they wore, they huddled in their sodden tents and grumbled. The men had hoped to return to India with money saved from their wages; now they worried whether they would ever live to see their families again.

Whitehouse, too, suffered a spell of malaria and was weakened by the disease. When, at last, the rains subsided and the wearied crews struggled back to work, they advanced only at a slow rate. Construction fell far behind schedule. According to the Railway's historian, Mervyn Hill, "It seemed as if the very spirit of Africa resented the intrusion of the white man's railway."

Tracklaying was also held up by a lack of locomotives and supplies caused by strikes in England. In March, the railhead was only at Mile 38½. Fifteen months had passed, and the railway was less than halfway to Voi, its first goal. The railway's backers worried. Would Parliament continue to support it? To soften expected criticism, the Railway Committee decided to claim in its report to Parliament that sixty miles of track had been laid instead of thirty-eight and a half.

If stubborn bush, heavy rains, malaria, and inadequate supplies were not sufficient obstacles, construction now suffered yet another blow. Early in April, dreaded bubonic plague broke out in Bombay, India. A quarantine was imposed; laborers desperately needed to reinforce the ailing railway work gangs were forbidden to leave India. The shortage of able workers lasted until late in the year.

However, despite the problems, tracklaying went well during April. Construction reached the eastern edge of the Taru Desert; the railway was at Mile 51, halfway to Voi at last. At this point, Whitehouse gained a valuable field assistant. Ronald Preston, a thin, dark-haired, handsome man, about thirty, arrived in East Africa from India, where he had built railroads and harbors. Preston came with his wife, Florence. He settled her in Mombasa and then went on alone to the railhead to take charge of tracklaying.

Preston had hardly begun work when torrential rains resumed. In May, more than twenty-four inches, a record rainfall for that month, washed away embankments and undermined and lifted crossties. Sections of track floated. At a meeting aboard his headquarters car, Whitehouse decided that the work crews should be turned back toward Mombasa to repair the damage caused by the rain. He directed Preston to rebuild bridges and embankments and realign the tracks before laying more track. Only after twenty-three days of reconstruction did the railway move forward.

The delays worried Whitehouse, but did not concern his work gangs. The laborers received their meager monthly wages whether or not they accomplished much; there was little to motivate them to work harder. Many loafed or pretended to be ill. Preston introduced a change in policy; wages were to be tied to the "piece," or actual work done.

The heavy spring rains that had so upset the construction schedule came to an end. Tracks advanced into the Taru Desert. Work gangs no longer slogged through mud in soaking rags. At last they were dry. But their relief was short-lived. The Taru is an awful place, a desert without sand or dunes. The red earth is arid and lifeless except for spikey bushes and a few trees. Africans do not graze their cattle there—the herds would die—and even wild animals stay out of the waterless Taru.

Trading caravans dreaded the blazing desert. The explorer Joseph Thomson left this vivid description: "The porter, wearied with a long march, and parched for want of water, presses on panting and perspiring under a broiling sun, made worse by the glaring red soil which reflects the rays as though they came from the mouth of a furnace."

After months of drenching rains, the work crews now longed for water. To aggravate matters, water trains from Mombasa occasionally were delayed or derailed, spilling their precious cargoes. Sometimes parched workers became so desperate they offered a whole month's wages for a drink. The sun blazing overhead, the searing steel rails, the blistering hot soil underfoot, and the choking, burning clouds of red dust made it seem as though the railway had entered Hell itself.

The Taru held yet another deadly menace. Part of a belt that runs some two hundred miles inland from the coast, much of the desert is "under the fly." That is, it is infested with tsetse (pronounced tet′ zee) flies. Tsetse flies carry sleeping sickness to human beings and to animals, and its prevalence in the area is another reason Africans avoid the Taru. The absence of sufficient local people who were willing to work as porters made it necessary to bring in bullocks, mules, donkeys, horses, and camels to carry supplies. But the tsetse fly and the heat killed the animals in droves.

At Mile 65, twenty miles into the Taru, the scene at the railhead resembled the aftermath of war. Fifty men had died from disease and the heat, and hundreds of others had been sent back to Mombasa seriously ill. Whitehouse, pale and weak from malaria, remained aboard his car. The smell of burning flesh hung heavily over the camp as fires consumed the carcasses of dead animals. Laying track across the Taru should have taken forty days, but because of the awful conditions the task took three months.

One evening a caravan of some thousand slaves traveling eastward chanced to stop nearby. It was led by the prosperous Arab slave trader, Saleh bin Amir. Saleh was a Muslim; he obeyed the holy Koran, his bible, which does not condemn taking non-Muslims as slaves.

Saleh came to visit the rail camp and had a long conversation with George Whitehouse and his staff. The wizened old slaver foresaw that the coming of the railway doomed his trade. Britain, he suggested, was determined to build a railway to end the slave trade only because it had not shared in the profits. Saleh was resigned, however, for he was nearing the end of his career, and he continued peacefully on his way.

Not until September, 1897, did the railway finally reach Voi. A less determined man than Whitehouse, having endured the hardships of the first hundred miles, might have paused at this milestone to rest himself and his men. But Parliament was concerned about the cost of the railway, and the Railway Committee sent a directive to Whitehouse urging faster progress. He was ordered to lay temporary tracks, bypassing rock formations, gullies, and other natural obstacles. When doing so was not possible, he was to improvise makeshift bridges. Getting the rails down quickly was the first consideration. Permanent works would come later. Whitehouse, committed by his orders to building the railway as quickly as possible, urged the crews on. Seemingly, the worst of the task was now behind him.

5

A Colonel Arrives

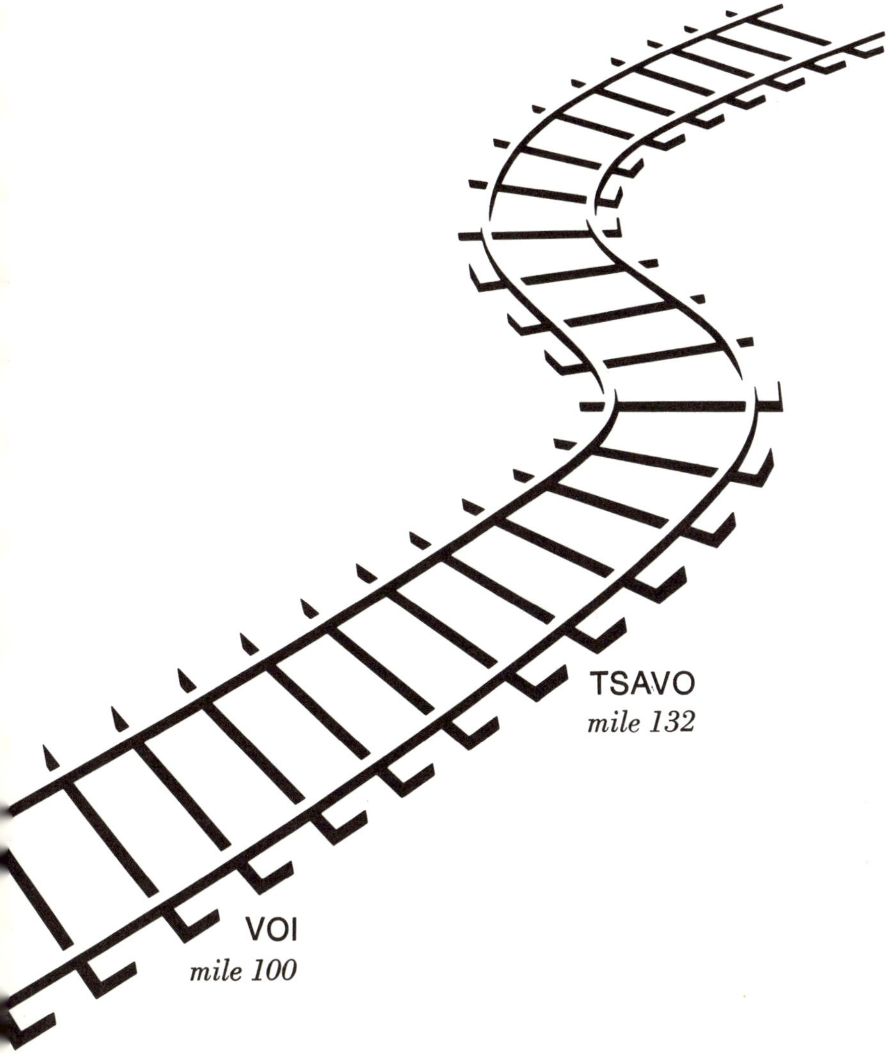

Tsavo is a forbidding wilderness. In contrast to the lifeless Taru, Tsavo is dense scrubland covered by prickly undergrowth and gray, leafless, stunted trees. Steep hills and blistered, rocky outcroppings rise from its red soil, hinting at past volcanic upheavals. From the foot of snowcapped Mount Kilimanjaro to the west, the Tsavo River cuts through this bushland, wending its way eastward toward the sea. A broad, swift-flowing river, its banks of lush vegetation are an oasis in the surrounding brush. Tsavo is the land of elephant, leopard, rhinoceros, hippopotamus, hyena—and lion.

By March, 1898, Whitehouse, Preston, and the crews had reached Tsavo. Three thousand workmen were encamped along both sides of the river, happy for the chance to wash away the dust of the Taru in the clear, clean water. A temporary bridge over the river had been built, but a permanent one was needed to support the heavy supply trains. Preston fully expected the responsibility for this project, so he was angry when the Foreign Office in London summoned Lieutenant Colonel J. H. Patterson, an experienced Army engineer, from India to build the new bridge.

Colonel Patterson arrived at the railhead, Mile 132, on March 8, at sundown. He was tall, slender, and the very essence of a proper British officer—formal and most conscious of his rank. He presented himself to Whitehouse in spotless khaki with a sun helmet, leather gloves, and shining, high, black boots. Then he retired for the evening.

Early next day Patterson inspected the camp. He saw that his belongings were brought to a suitably large tent, which would be his quarters, and he assembled the staff that would attend to his personal needs: four servants, a cook, and a *jemadar*, or headman. The *jemadar*, a huge, turbaned Indian Sikh named Ungan Singh, was to see that Patterson's needs were satisfied. The colonel was a finicky man of strict habits. He brought into the wilderness of Tsavo a collection of costly hunting guns and an assortment of personal effects, including his own rubber bathtub.

The next thing Colonel Patterson did was send for Ronald Preston. When Preston entered the colonel's tent, Patterson was examining his prized guns. Preston watched in silence as the colonel made no move to greet him. This treatment demonstrating the colonel's superior position offended Preston. Though both men had come from India, their backgrounds were different. Patterson was born into wealth and privilege; Preston, an orphan, had risen from humble circumstances.

Finally Preston became so impatient that he laughed at the pompous colonel. Patterson, furious but in control, turned and coldly acknowledged the younger man. From that time on they were enemies.

Colonel Patterson was noticeably different from the other railway engineers. Not only his military bearing and superior attitude, but his personal ways as well made him conspicuous. Whereas most of the staff adjusted to construction-camp life, Patterson made few compromises, even deep in the East African bush. Each day at dawn the colonel drank a pint of hot tea, followed shortly by his breakfast, which consisted of a tray of cold meats. Afterward Patterson took a warm bath in his own rubber tub. He scrubbed himself; then a board was placed across the tub. On this shelf was a mirror, a razor, and the Holy Scriptures. Each morning, as he sat soaking, Colonel Patterson shaved and read the Bible.

Normally, Patterson's bath was prepared by his lesser servants. One day, however, these servants were building a fence on which the colonel could display his hunting trophies, and the job of preparing the bath fell to Ungan Singh, the colonel's headman. Though the proud Sikh was not familiar with the routine, he did not hesitate. Singh gathered firewood and arranged it beneath the rubber tub. Next he poured a bucket of cold water into the bath, lit the firewood, and raced to the river for more water. Minutes later a column of black smoke and a strange odor signaled that something was wrong. Singh rushed back to the bath. The bottom of the sahib's prized rubber tub had melted and fallen into the fire. Though the colonel must have been upset, he managed somehow to get along.

Patterson had been at Tsavo only a short time when reports of lions dragging off workers in the night reached him. Preston had discovered a laborer who had failed to report for work one morning down by the river. His body was severely mutilated—the head and feet were spared, but the rest of the flesh had been torn away. A few days later a second victim was discovered on the riverbank. This time even the flesh of the face had been ripped away, leaving the skull grinning grotesquely.

Though Patterson was told that pug marks, the pawprints of a lion, had been found near the mangled corpses, the arrogant colonel doubted the evidence. He suspected that other laborers had murdered the victims and stolen their money. He was convinced that the workers were not above killing each other and using pronged ground-clearing tools to imitate lions' footprints.

Patterson was too busy to investigate the matter further. The new bridge took all his attention. Stone piers were to be built to support a three-hundred-foot iron span. Workmen and supplies continued to arrive. The area resounded with the sounds of hammers, drilling, and blasting.

Three busy weeks sped by. Then one morning a terrified workman awakened Patterson before dawn. A lion had come through the open door of the tent he and four others shared with Ungan Singh. The *jemadar*, large as he was, had been seized and carried off by the beast.

Patterson began an immediate search for Singh. There was evidence of a fierce struggle and telltale pug marks. Patterson finally believed that lions were raiding the camp. To his horror, he discovered a trail of blood and the heel marks of a man being dragged. Then he came upon poor Singh; the remains of his body were torn into bits of flesh, bone, and blood. ". . . the unfortunate

jemadar's head had been left intact save for the holes made by the lions' tusks on seizing him, and lay a short distance away from the other remains, the eyes staring wide open with a startled, horrified look in them. . . . It was the most gruesome sight I had ever seen," recorded Patterson in a book he later wrote.

Lions do not usually attack people; the human scent is repugnant to them. But older or injured lions, too slow to catch other prey, may sometimes set upon human beings. They kill their victims with a blow of a powerful front paw and then tear off the skin to drink the fresh blood. Signs showed that two lions had fought for poor Singh's body. Patterson, ever the man of honor, vowed to avenge his loyal *jemadar* and to protect his workmen. The railway had encountered yet another obstacle—man-eating lions.

6

The Man-Eating Lions of Tsavo

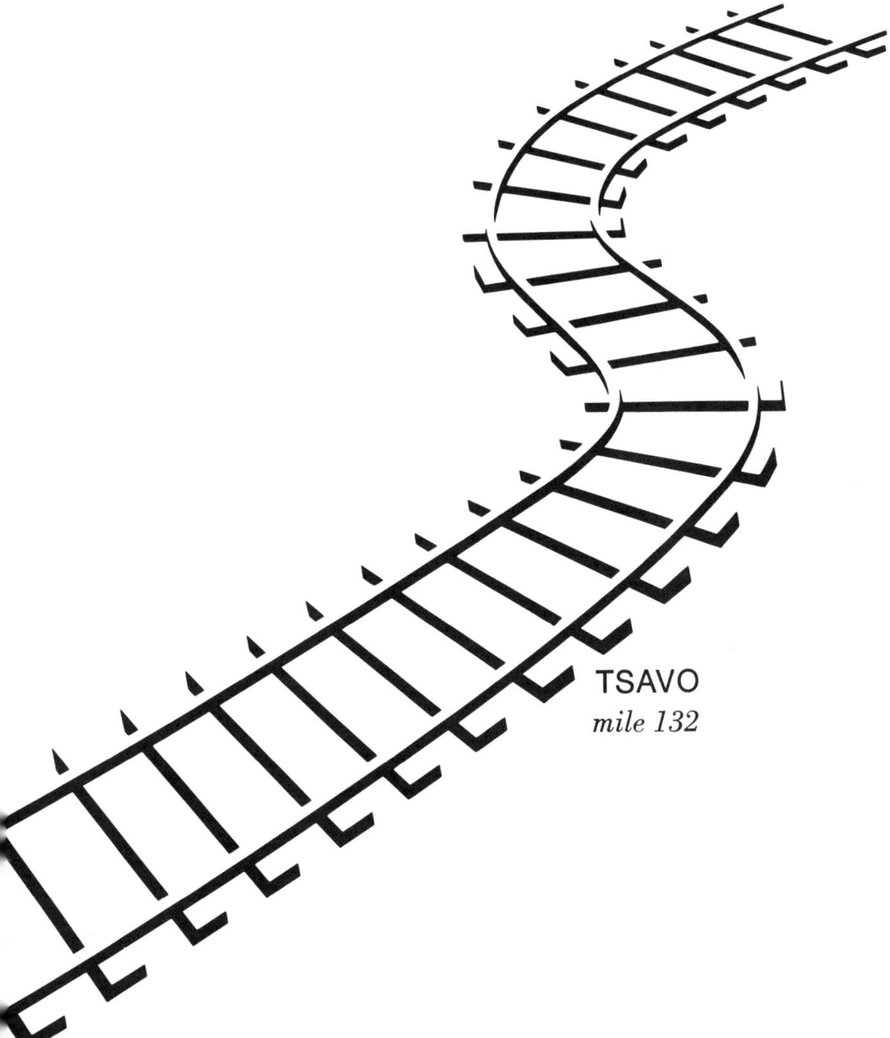

The horrible death of Ungan Singh spread fear through the camp. The defenseless laborers were terrified by the man-eating lions. They built *bomas*, barricades of prickly thorn, around their tents, and word was passed that henceforth tent doors were to be kept closed at night.

That evening, when the uneasy camp turned in, Patterson took up watch with a rifle and shotgun in a perch in a tree near Singh's tent. He expected the lions would return to the same spot for another victim. Several frightened workmen saw him in his perch and climbed up onto limbs around him, hoping to find safety off the ground.

Soon Patterson heard lions roar in the distance. They seemed to be approaching. Then silence. Patterson made ready; lions become absolutely still before they strike. Suddenly shrieks of terror from another part of the camp ripped through the dark; the lions had claimed a new victim.

The next morning Patterson learned that the latest attack had occurred a half mile from where he kept his vigil. As the laborers' tents spread for eight miles on either side of the river, the colonel realized that he could not possibly protect the whole camp from any one point.

Patterson tried another tactic. He staked out a goat, hoping to bait the lions. Again he climbed into a tree with his guns and waited. Once more, however, distant screams in the night told him that the lions had struck elsewhere.

Again and again Patterson matched wits with the lions. They seemed to sense wherever he was, however. Each time they eluded him, choosing to attack elsewhere, no matter how carefully Patterson concealed himself. But the lions did not always succeed in getting a victim. Once, for example, a stunned worker was left behind while a lion dragged off his mattress. On another occasion, a hasty lion ignored a horrified laborer and made off instead with a very large bag of rice.

The camp was not raided every night; days would sometimes pass without incident, raising hopes that the beasts had moved on. But the lions spared the work camp only to attack the villages of the Kamba tribe nearby. They always returned.

Progress on the railway was affected; the workers were near panic. While Patterson supervised the building of the bridge on one side of the river, tracklaying shifted to the other bank. Preston, in charge of this activity, cleverly played on the crews' fears to hurry their work. He convinced the laborers that the lions lurked by the river and that by working faster they would get away and be safer sooner. By the end of the month, Preston and his fear-driven gangs had moved on westward. Patterson remained at the river with some five hundred men to complete the permanent bridge.

The workmen left behind felt even less secure. They moved their tents closer together, built higher barricades, and reinforced them. Each night guards were posted inside the *bomas*. They kept fires burning and made noise to scare off the lions. Empty oil cans, suspended from trees, were rigged so that watchmen, sitting safely within their tents, could create a clamor by pulling on ropes. Despite these measures, the determined man-eaters often broke through into the camps. Workers' tents—even the hospital—continued to be raided.

By mid-April, the lions, emboldened by their success, dared to appear in daylight. Now Patterson hunted them night and day. He stalked them out in the bush; more than once he trapped himself in such dense, thorny brush that his gun bearer had to disentangle him.

In spite of the lions, the work at Tsavo somehow progressed. Permanent tracks replaced temporary ones, but construction of the bridge was held up for lack of hard stone to build the supporting piers. After so much misfortune, Patterson finally had a stroke of luck. Miraculously he found the right kind of stone a few miles away. It was too far to be carried to the construction site, but the ingenious engineer devised a trolley to transport it. However, by August, when the bridge should have been finished, it was still under construction.

Added to all its other woes, the Uganda Railway now had difficulties with the workmen. Patterson soon discovered that a number of his stonemasons were unskilled frauds. The pretenders had passed themselves off as masons to collect higher wages, but they could not do the work. When the impostors were exposed, they became rebellious and rallied other workers behind them. One morning, tired from a night-long watch for lions, Patterson decided to stop at the stone quarry on his way back to camp instead of visiting it later in the morning as was his custom. He found the quarry unusually quiet. Instead of working, the men were snoozing under trees and playing cards. The outraged colonel scolded and fined them. The shirkers became enraged at Patterson and plotted to kill him.

An informer warned Patterson the night before the attack was to take place. He revealed that the conspirators planned to trap the colonel in the narrow part of the quarry, kill him, and hide his body in the bush where the animals would eat it, making the murder look like lions' work. Undismayed, Patterson visited the quarry as usual the next morning. His arrogance was such that he could not believe that the workmen were capable of planning or carrying out such an act. Any doubts he might have had about the plot vanished as he walked deeper into the quarry. Sure enough, a foreman led him to the narrow place he had been warned about.

The trap was sprung. Patterson found himself surrounded by a large band of cutthroats wielding crowbars and hammers. An attacker seized his wrists, but the colonel wrenched free. Another lunged at him as the agile Patterson stepped aside, causing the thug to fall. The attack stalled for a moment, long enough for Patterson to climb atop a boulder.

From this position, the colonel summoned up all of his authority and proceeded to give the workers a tongue-lashing in Hindustani, their own language. He boasted that he had known of their scheme and that his death would never be attributed to lions. They would hang for his murder. At the same time, he offered not to report them if they returned to their work. The mutineers, taken aback by the colonel's bold speech, disbanded and returned glumly to their tasks.

Patterson's problems were not over, however. The troublemakers schemed again to murder him that very night. Once more an informer told Patterson, and the colonel was able to telegraph for assistance. The band of mutineers was arrested and brought to trial in Mombasa.

Peace returned to Tsavo—even the lions were quiet—and the remaining weeks of summer passed without incident. But then in the final months of the year the lions returned with a vengeance. Terror ruled Tsavo. The weary, defenseless laborers lay sleepless on their cots, unable to ignore the danger that lurked around them. "The workmen," according to Patterson, "firmly believed that they (the lions) were not real animals at all, but devils in lions' shape. . . . They were quite convinced that the angry spirits of two departed native chiefs had taken this form in order to protest against a railway being made through their country. . . ."

On December 1, all but about fifty loyal workers deserted Tsavo en masse, climbing aboard a supply train headed for Mombasa. All work ceased.

Then, on the morning of December 9, Patterson was summoned with the cry, "*Simba! Simba!*" ("Lion! Lion!" in Swahili). He grabbed a two-barreled rifle lent him by a police officer and stalked the beast until it stopped and turned to face him. Patterson had only a moment to aim and shoot. As fate would have it, the borrowed weapon was defective. The right barrel stuck and didn't fire. Patterson was unnerved, and momentarily he forgot that the weapon had another barrel. Just in time, he remembered and pulled the second trigger. The rifle fired, wounding the lion and sending it off into the bush.

That afternoon Patterson had a platform built. It was little more than a flimsy scaffold atop four poles, twelve feet above the ground. Night fell. A donkey was set out nearby as bait. Patterson waited alone on the rickety platform in the darkness. Time passed slowly. Then a twig snapped. Patterson came alert. He heard a low growl and knew his enemy had returned.

The lion, ignoring the donkey, began to stalk Patterson. The hunter himself had become the bait. For the next two hours, the lion prowled stealthily around Patterson's frail perch. Closer and closer it came, all the while concealed from view in the brush. Patterson's nerves were stretched to the limit. He dared not move. He held himself so still, in fact, that an owl mistook his head for a stump and flew into him. Startled, Patterson nearly fell off the platform. He held on, however, and his patience was rewarded; the lion stalked into the clearing. Patterson took aim and fired. His shot struck the lion, wounding it. The powerful beast roared and plunged into the concealing bush, but Patterson kept firing and killed it at last.

The next morning eight men were needed to carry the man-eater into camp. It was an extraordinary beast, almost ten feet long. The workers celebrated the kill, and many Africans came to see the lion and to praise Patterson. Within the month, Patterson killed the other giant lion that had terrorized the camp.

The incredible man-eating lions of Tsavo had eaten twenty-eight Indian workmen and scores of Africans. Word of the deaths of these beasts brought back the laborers who had deserted earlier. The bridge, now seven months late, was finally completed, and on February 7, 1899, the first train crossed it. Two days later a fierce storm struck. To Patterson's satisfaction, the new bridge withstood the test perfectly.

7

The Shortcut

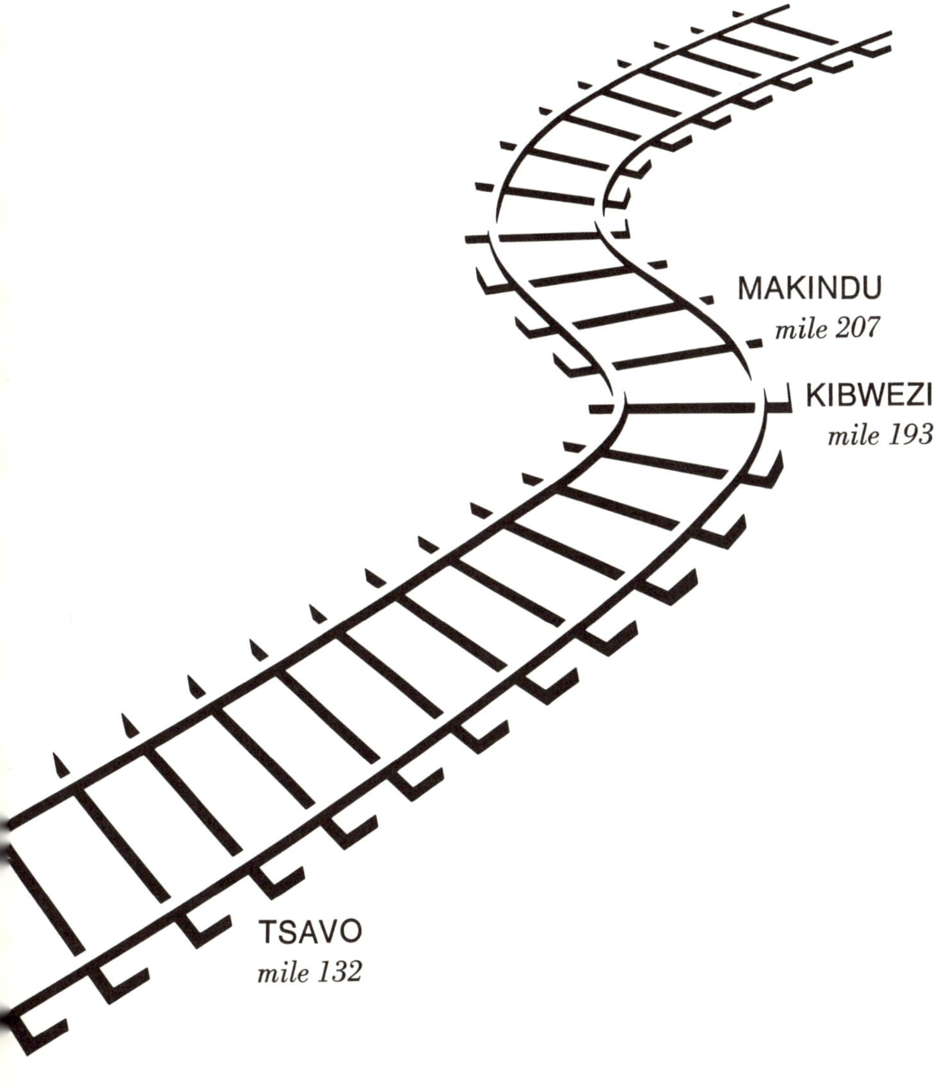

While Colonel Patterson and his terrorized construction gangs labored to finish the Tsavo bridge, Preston and his tracklayers had pushed onward. In three months—by July, 1898—they had managed sixty miles, from Tsavo to Kibwezi, Mile 193. The tracklayers, whose numbers had swelled to 7000, had encountered difficulties because of heavy rains, dense bush, and continued attacks by lions. Though progress had fallen short of the mile a day Preston hoped for, he was pleased at least to have left behind the overbearing Colonel Patterson.

Also, Ronald Preston was happy to have his wife, Florence, rejoin him. Married seven years, the Prestons had two young children, who were being cared for back in India. Florence, a determined woman who did not shrink from hardship, had come to Africa to be with her husband and only extreme danger had parted them. Now she planned to stay with him until the railway was completed.

The station at Kibwezi, where the Prestons were reunited, was little more than a stop where the train from Mombasa could take on fuel and water for the trip back. A Scottish missionary settlement was nearby. The countryside was open and green, and the air had a refreshing crispness (Kibwezi is 3000 feet above sea level). Yet the cool, clean breezes that welcomed Florence Preston were deceptive. They carried mosquitoes bearing blackwater fever, a dread form of malaria that blackens the urine of its unfortunate, fevered victims. The disease had so devastated local people that Kibwezi had a reputation as the home of blackwater fever. A third of Preston's work force was already disabled by this newest plague.

Preston's weakened gangs continued to push forward slowly, leaving the tsetse-fly belt behind. "Under the fly," the pace of construction had been held back by problems in moving supplies to the work sites. Some 1850 animals had been brought in to help, but about five out of six had died. Now steam-driven tractors arrived by train from Mombasa to pull the wagonloads of supplies to the tracklayers.

With this new help, the railhead advanced to Makindu, at Mile 207, by September. Makindu is a place known for its dom palm trees, which are remarkable for their symmetrical branches. They flourished at Makindu, where their roots could obtain groundwater from a stream. In this scenic place half the camp came down with chills and fever. The railway's advance slowed to a crawl.

With construction all but halted, Ronald and Florence Preston took time to hunt. The cook welcomed the fresh meat they provided. They collected kudu horns and other antelope trophies as well as hides to adorn their tent. Florence Preston did what she could to make the tent a home. She covered a table—a slab of wood on two sawhorses—with a clean, fringed cloth. She arranged the simple furniture—a bed, some chairs, and a chest—and gathered wild flowers. She washed dishes and clothes in a nearby stream. Often she sat at the tent's entrance in a properly long black skirt, a full-sleeved blouse, and a brimmed black hat, knitting or embroidering. A wild gray parrot she had befriended sometimes perched on her shoulder.

Other wives also tried to live with their engineer husbands at the rail camps. Many, however, found the hardships too severe and gave up to return to the coast or to go home. The Prestons' simple tent with its homey touches and ready welcome for tea and conversation was a haven for lonesome engineers.

Meanwhile, back at the depot in Mombasa, George Whitehouse received cable after cable urging, then demanding, faster progress. Nearly three years had passed, and construction of the Uganda Railway was only one third completed. Given the obstacles—parching desert, deluges, dense thornbush, disease, and man-eating lions—such progress was remarkable. But Parliament and a Railway Committee wary of criticism were dissatisfied. London

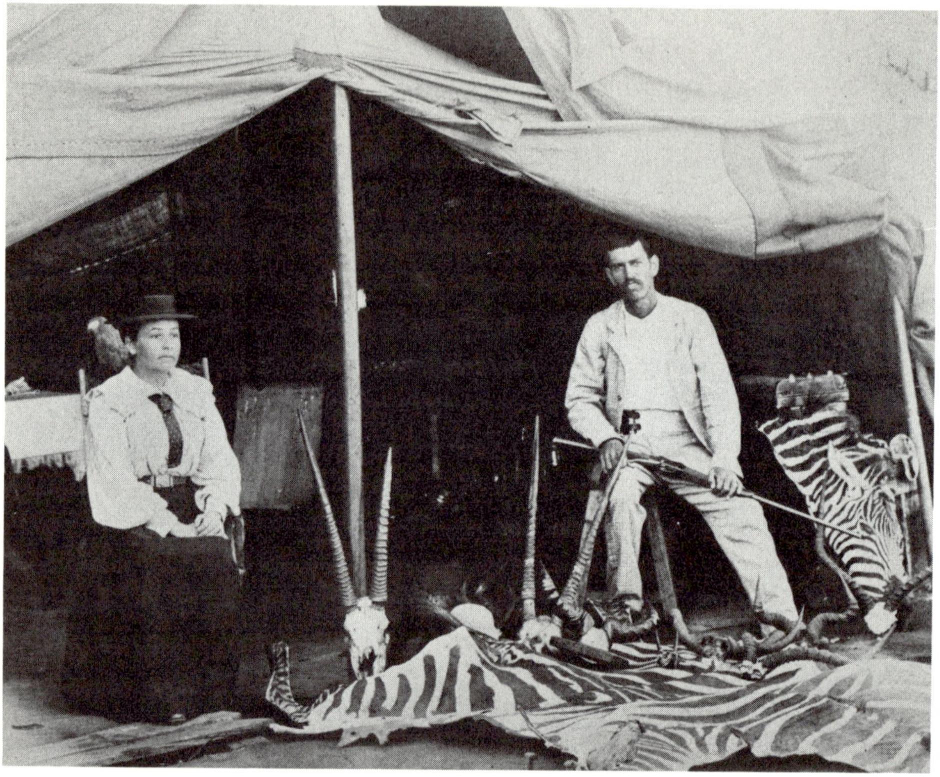

seemed unaware that each costly mile of track also exacted a price in human life. Already three hundred men had died and seven hundred had been sent home, disabled by injury or disease.

George Whitehouse was warned that as chief engineer he had the responsibility to speed construction and cut costs. The conscientious engineer needed no reminder of his duty. He made constant efforts to supply men and materials to Patterson, building the bridge at Tsavo, and to Preston, laying tracks near Makindu.

At this time exciting word came from surveyors plotting the railway's route. They had discovered a shortcut between the Rift Valley and Lake Victoria, one that could save time and money. Perhaps eager to escape, even temporarily, from the terrible demands at his headquarters, Whitehouse decided to go personally to investigate the new route.

He embarked by train with a caravan of porters from Mombasa to the railhead near Makindu. From there he and his porters set out on foot. They walked all day and camped at night. The caravan crossed great, open plains and slogged through marshes. About halfway between Mombasa and Lake Victoria, just before the highlands begin, they came to a caravan trading place called Nairobi, or "cold water" in Maasai, because of a cool stream that runs through there.

Whitehouse received a confirmation in Nairobi that surveyors had checked and confirmed the reported shortcut. Eager to report good news at last, he hastily sent word of the discovery to London from a telegraph station set up in advance of the railway. He might well have turned back at this point—his expedition no longer was required—but Whitehouse was anxious to check the shortcut himself.

Ronald and Florence Preston sit in front of their tent. A parrot perches on Mrs. Preston's shoulder. The other photograph shows George Whitehouse (with dog) and senior staff in Nairobi.

The caravan reached the Great Rift Valley. While descending its eastern wall, Whitehouse surveyed the steep slope and measured the grade the railway would have to overcome. The caravan, now on the floor of the valley, crossed a stream that marked the boundary between the East Africa Protectorate and Uganda at that time and then turned northward. Whitehouse halted his advance at Lake Naivasha. There he paused to admire the beautiful birdlife—the cormorants, ibises, and flocks of pelicans—and took time to write a long letter to London, telling about the shortcut in greater detail and reporting that he was on his way to investigate it. He sent the letter by a relay of runners to the railhead, where it would travel by train to Mombasa and then by ship to England.

The caravan moved northward to Lake Nakuru, and Whitehouse marveled at the sight of thousands of magnificent flamingos. At this point he made a decision to compare the two routes. He would take the original course to Lake Victoria and then return by the new, shorter one. By taking the longer route first, he could also arrange to meet his younger brother, a naval lieutenant, assigned to a military outpost at Port Victoria in the northeastern corner of the lake. Accordingly, the caravan continued past Nakuru before turning west to climb up out of the Rift Valley and trek on to the lake. The journey from Mombasa had taken thirty-five days.

Whitehouse was delighted to see Lake Victoria, the railway's destination. There the famous African explorers, John Speke and James Grant, had come to seek the headwaters of the Nile. This great lake, the largest in Africa, is as big as all of Scotland.

At a creek near Port Victoria, Whitehouse met his brother. He sent his porters ahead on foot, telling them he would meet the caravan in a week at a place fifty miles farther to the southeast, where the new route would reach the lake. Meanwhile, the chief engineer visited with his brother, exchanged family news, and relaxed. The younger Whitehouse, also an engineer, was beginning a marine survey of the lake, and George Whitehouse accompanied him while he made measurements of the Port Victoria area by boat.

When Whitehouse took leave of his brother and met his caravan at the proposed end of the new route, he discovered a problem engineers would have to solve once the railway was completed. A harbor was to be built to carry trade to and from central Africa, but Whitehouse realized that the water was too shallow at the intended site to accommodate the large steamships expected.

8

To Kikuyuland

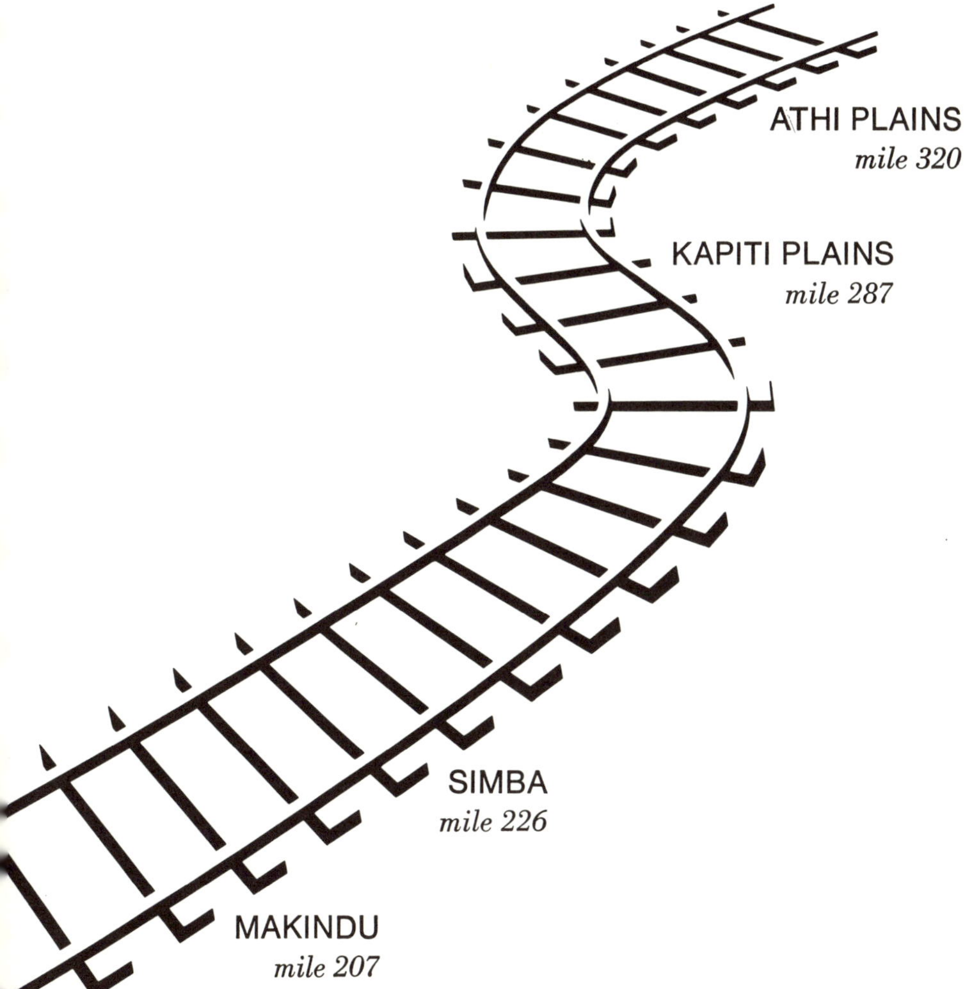

During Whitehouse's absence, Preston and the tracklayers advanced only nineteen miles to Simba. Again malaria was the main problem. No one had been spared, not even Preston. Fever and weakness had forced him to spend days in his tent under his wife's care. The entire work force had been replaced with fresh crews from Mombasa.

Also, the weather, which had hindered the railway from the outset, continued its mischief. While a drought prevailed in the general region, heavy, local storms repeatedly struck the railway and flooded the tracks. All traffic and work were halted for two weeks.

One day, after it had poured steadily for four hours, the rains suddenly ceased. Relieved, Florence Preston came out from her sodden tent. The camp was astir with the sounds of work crews gathering. Suddenly there were screams. All along the embankment, thousands of small, terrifying creatures were emerging from holes and cracks in the earth. They rose, arcing their jointed, poisonous tails over their bodies. Scorpions!

Near panic prevailed; work was forgotten. Frenzied laborers jumped and struck the ground repeatedly with their tools. The scene resembled a madmen's dance. Finally braver men noticed that the half-drowned scorpions were waterlogged and sluggish. Boldly they scooped the creatures into pails and encouraged other workers to help. The scorpions were dumped onto fires. Order was restored.

Shortly afterward, Whitehouse arrived at Simba, Mile 226, exhausted from his long trek. On the journey back from the lake, he had explored the new short route and found it suitable for the railway. Altogether he had been gone two months and had covered eight hundred miles. To Preston and others at Simba, Whitehouse's report that the shortcut would save an estimated 114 miles (actually it saved 75) was good news. Unfortunately, they had the less pleasant task of explaining why so little progress was being made.

Seemingly all of nature—weather, disease, death, creatures large and small—was conspiring against the railroad. Wooden railway crossties treated with a protective coating of creosote to resist insects were used on sections of the line in place of steel ties. But white ants defied this sticky barrier. They invaded the crossties and weakened them, causing derailments. When gray guinea fowl, the size of small chickens, came to feed on the troublesome white ants, they apparently were helping the railway. But the birds dug so many holes that an embankment collapsed, forcing an engine from the tracks. Even small caterpillars and swarms of locusts hindered trains. Thousands of their little bodies crushed by the huge locomotives made the rails so slippery that traction was reduced. As a result, a train laden with supplies derailed.

Larger animals, too, continued to threaten the railway. Lions prowled the tracks. Traffic was held up because signalmen at Simba station were too terrified to venture out at sunset to light the signal lamps. Workers panicked when rhinoceros appeared at a site. Hours were lost while men hid and waited for them to leave. One of these brutish animals lifted a hospital tent on his horn and dragged it off several hundred yards into the bush.

Animals interfered with telegraph communications as well. The rhinos were fond of rubbing their hides against telegraph poles, often toppling them over. Giraffes frequently bumped their long necks into the telegraph lines, acquiring necklaces of wire, and baboons seemed to take pleasure in swinging from the overhead lines.

Despite the obstacles, however, the railway pushed northwestward in the final months of 1898. It advanced across savannas, great expanses of flat, grassy plains. Except for a few umbrella-shaped acacias, the grasslands were devoid of trees. On these Kapiti and Athi Plains grazers such as the wildebeeste, zebra, gazelle, impala, and warthog feed while predators like the lion and leopard stalk them. Buzzards and marabou storks wait in perches atop the few acacia trees to feast on the remains of kills.

Tracklaying progressed across the plains during the first months of the new year, 1899. These times were difficult for the Africans along the railway's path. Rinderpest, a deadly cattle disease, had killed herds and much of the game in the region known as Ukamba Province, which included adjacent Kikuyuland to the north and west. The nomadic Maasai, whose way of life and survival depended largely on their cattle, had moved their precious herds south to safety. The Kamba and Kikuyu, however, saw their herds dwindle. These people, especially the Kikuyu, were agricultural. They might have had enough food from the crops they raised, but a severe drought hit their districts at about the same time.

The spring rains failed; maize and millet, the staples of the Kamba, withered. The plight of the Kikuyu was less severe at first. The rich, red soil of their little farms produced sugarcane, bananas, maize, sweet potatoes, beans, and millet. Kikuyu farms were scattered on high, hilly ridges, which received some moisture. Some sections were spared entirely during the early months of the drought. But as the year wore on and the drought continued, Kikuyu crops suffered too, and increasing numbers of Kikuyu as well as Kamba abandoned their homesteads to search for food. Their misery was increased by a severe epidemic of smallpox.

The railway, which many in Parliament had promised would benefit the Africans, now had a chance to do so. Extra provisions of flour, rice, fruits, and vegetables were sent up from Mombasa to feed the starving Africans. Women and children came begging to the trains. At first they arrived in small numbers, then in swarms. They lined the track embankments hoping for food, an army of withered, emaciated, living skeletons.

When the trains came, a hellish scramble took place. People climbed over one another, pushing, clawing, desperately reaching for food. There was never enough to feed them all. When the trains moved on, starving crowds were left behind. Some followed the tracks, hoping to come upon another train. Others remained quietly, lacking even the strength to cry out. An Englishwoman, the wife of a missionary, witnessed the horror and wrote: "No

matter where one went corpses strewed the tracks. Little skeleton babies were found crying by the dead bodies of their mothers."

The rescue efforts slowed construction of the railway; supply trains bearing food could not carry tracks. Whitehouse was moved by the plight of the Africans, but he could not feed his 13,000 workmen, the starving local people, and continue to build the railroad. He appealed to his government to take over the relief effort. The food trains were stopped, while the government made plans to create food stations, but conditions grew worse. Desperate, starving Kamba tribesmen began to raid the construction camps for flour. They crept up at night, shooting poisoned arrows. Many laborers died.

Finally food stations located near the railway were set up in the famine area. Communications were so poor, however, that news of the new relief efforts often did not reach stricken areas. Thus, many who might have been aided never learned that help was nearby. Others heard too late; they were already too weak and sick to travel. An estimated 25,000 Africans died in the famine.

Despite the misery in the region, the railroad advanced to the outskirts of Kikuyuland. Although the Kikuyu had long tolerated and even profited from the trading caravans that had skirted their lands, the railway seemed to worry them. Great numbers of strangers, led by white men, were approaching their land. The Kikuyu were mindful of the ancient medicine man's prophecy. Had not a famine and white-skinned strangers come as predicted? And now the railway—an iron snake—was coming too.

The Kikuyu were determined to keep white men from entering their territory. But the pale-skinned strangers, whose railway would run only around the fringes of Kikuyuland, were equally anxious not to antagonize the Kikuyu and risk delay. So though minor raids on railway gangs did occur, the Kikuyu made no major assault against the railway.

9

A Masterpiece of Engineering

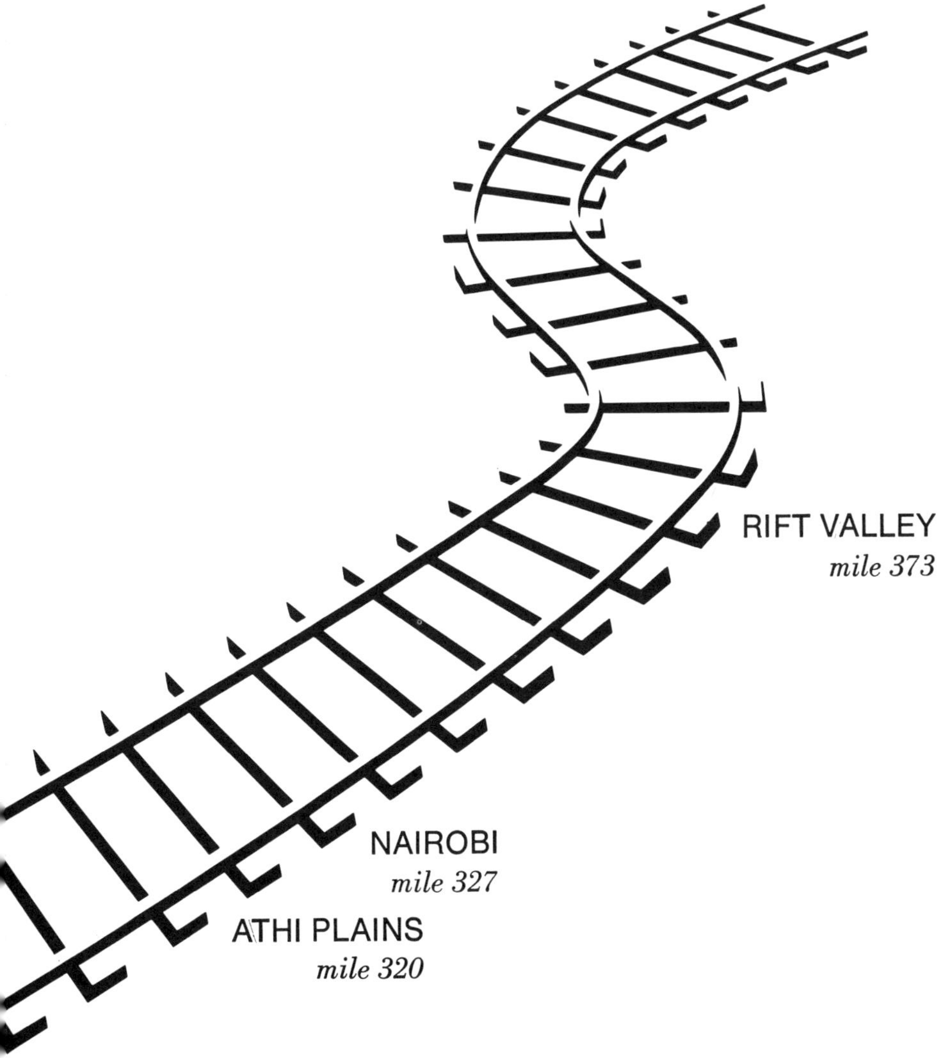

When Colonel Patterson and his labor force had completed the bridge at Tsavo, they moved onward. They improved or replaced the makeshift bridges Preston and the tracklayers had hastily thrown up, and they also constructed a series of stations.

Patterson and his gangs had finished building a bridge across the dry bed of the Athi River when Whitehouse called him to the railhead at Nairobi, Mile 327. The tracklayers were already encamped there. Patterson was given the task of transforming this bleak, swampy place into a major new depot for the railway. Nairobi was chosen as the new headquarters because of its location midway on the route between Mombasa and Lake Victoria. Also it was situated conveniently on the last stretch of flat land before the Rift Valley.

Patterson arrived at Nairobi at midday, June 5, 1899. Troops of masons and builders climbed down from the train and began to unload their equipment. Colonel Patterson had considerable personal property to be off-loaded. Numerous crates filled with his possessions were carried from the train, and his horse, a stallion with the impressive name Blazeaway, was led from a cargo car.

Patterson settled himself and his servants on a hill above—and apart from—the Prestons and the rest of the railway staff. His quarters were also farther from the nightly chorus of croaking frogs to which the railway community had to become accustomed. Patterson enjoyed his heroic reputation as the lion killer of Tsavo, and he behaved with even more arrogance than before. The haughty colonel now rode regally about on his horse, giving orders grandly. Within weeks, however, even Ronald Preston was impressed by the speed with which the colonel converted the barren marshland into a complex of roads, bridges, workshops, and administration buildings.

The main offices and the headquarters personnel of the railroad were removed from Mombasa and installed in the new depot. By August, 1899, a station had been built. It was a simple, wooden structure roofed over by strips of corrugated iron and divided into a cargo-and-storage area and a passenger platform.

Regular passenger service between Mombasa and Nairobi had started. The railway community quickly became used to the arrival of trains, but to the Africans in the area such occurrences were extraordinary events. Maasai, smeared with sheep's fat and red earth, naked except for goatskins draped over one shoulder, with ornaments in their ears, flocked to see the mighty fire snake pant and hiss.

Trains were used to transport British troops quickly inland to Uganda, where uprisings sometimes occurred. They now were also carrying raw hides, skins, ivory, and horns from Nairobi toward the coast for export. Rail freight already was proving to be faster, cheaper, and more reliable than caravans of human porters.

Nairobi's population grew rapidly in the months following the railroad's arrival. The governing officials of Ukamba Province and their staffs set up offices there. Many Indian laborers, having completed their three-year contracts with the railway, had chosen to remain in East Africa and had set up trading posts and shops at railway stations along the line. Now they came to Nairobi and established a bazaar. It consisted of many dimly lit, poorly ventilated little shops known as *dukas*.

The shopkeepers sold sacks of salt, beads, wire ornaments, blankets, cloth, and dried meats for rupees, since the railway used this Indian currency to pay the enormous work force brought to East Africa from India. Gradually, many Africans, too, began to accept wages paid in rupees instead of beads and wire. With these easily carried coins they could buy a variety of goods in the *dukas* or purchase goats. The railway was beginning to affect the age-old ways of Africans.

Nairobi's development brought with it serious health problems. The increasingly crowded railway hub had inadequate sanitation. Sewage did not drain away effectively from the swampy ground, especially when seasonal rains flooded the area. Garbage accumulated and rats flourished. These conditions brought diseases. Soon it was suspected that the bazaar harbored cases of plague.

Colonel Patterson was determined to investigate the rumors, and accompanied by Doctor Waters, the railway surgeon, he inspected the bazaar. The dingy, foul-smelling market offended the genteel colonel, but he plunged on. Suddenly they came upon a sick man lying at the back of a *duka*. Waters examined the stricken man; a festering sore and a rash, signs of plague, were present. Movement beneath the floor of the shop caused the physician to pry up a floorboard. He found what he suspected—rats. The finicky colonel watched in horror as the physician probed through a dead rat's fur and found fleas.

Patterson knew that fleas transmitted the dread disease from infected rats to human beings. Without troubling to consult Whitehouse or to obtain authorization from the colonial Government, Patterson sent notice to the shopkeepers that they had one hour to save what they could and clear out of the bazaar. The frightened merchants scrambled to remove what goods they could from their *dukas*. True to his word, sixty minutes later, Colonel Patterson returned and set the market afire.

The bazaar burst into flames and the rats poured out in living streams. Merchants watched helplessly as the blaze consumed their shops laden with goods. Colonel Patterson returned to his tent and watched the smoke rise; he was pleased with his efforts.

Whitehouse, however, was outraged. He summoned Patterson to his office and denounced his impulsive destruction of the bazaar. Whitehouse pointed out that the railway would now have to spend its precious funds to reimburse the shopkeepers for their ruined property. He dismissed Patterson's explanations. The rats, driven out from the bazaar, now infested other buildings. They would only return when a new market was built. Patterson listened to his chief patiently, unruffled. He continued to believe that he had spared Nairobi an epidemic.

Near the completion of his service, Patterson left for England at the end of the year. His battle with Nairobi's rats capped his career with the Uganda Railway.

Meanwhile, Preston and his gangs made their way easily through the cool, wooded highlands northwest of Nairobi. On October 2, 1899, they reached the Great Rift Valley. This extraordinary break in the planet's crust is thousands of miles long. It runs from Asia Minor through Africa into Mozambique.

Ronald Preston stood on the brink of the Rift, looking out into the hazy, purple distance. Never in his life had he seen anything so impressive. Great extinct volcanoes rose from the Rift's flat bottom, fifty miles wide where he stood. Before him the wall plunged sharply some 1500 feet. He must have wondered how he could build a railroad down this steep slope.

The best way to conquer the sharp descent seemed to be a northward detour. In this direction a series of hills descend gradually, like steps, from the top of the wall to the valley floor. The engineers planned a path ten miles long, which would curl through these hills. Eight bridges would be built to support the railway along this winding route.

Preparations for the descent began, and quantities of steel needed to build the bridges were sent for. At this time, however, Great Britain was fighting a war in South Africa (the Boer War, 1899–1902), and steel was diverted from the railway. Nevertheless, the pressures on Whitehouse to complete the railway did not lessen.

Now his engineering genius became apparent. Whitehouse put aside the plan to build the bridges and devised a temporary solution to get to the valley floor. Thus, construction could proceed across the valley without delay. The temporary system would go directly down the steep slope, taking advantage of a series of ledges to break the descent. Whitehouse designed special freight cars for the route. Their supporting undercarriages were higher at one end than at the other, enabling the cars to ride level on the sloping track. In this way they could carry loads of supplies down the inclines between the ledges. Cables tied to these wagons would pass around pulleys to empty cars on a nearby parallel set of tracks. As loaded wagons descended on one set of tracks, the empty cars would be lifted, like a counterweight, on the other. The design was a masterpiece of engineering.

Work had hardly begun on this ingenious plan when shortages of materials brought it to a standstill. Weeks dragged into months; Christmas came and passed with little cheer among the impatient English engineers. The old year was ending, and the small group sent to build this railway on the soil of East Africa, far from their homes, must have felt lonely. At midnight on December 31, Ronald and Florence Preston dressed for the occasion and joined other engineers and their wives on the eastern wall of the Great Rift to welcome in the new year of 1900. How out of place they must have seemed, those lonely strangers in their freshly pressed jackets and dresses, greeting the new century there.

In January the long drought that had parched the region for two years came to an end. Drenching rains fell once more. They turned the construction camp into a sea of mud and created serious problems back at the supply depot in Nairobi. The swampy land drained so poorly that the town flooded. The rains saturated the highly absorbent black soil of the Athi Plains east of Nairobi, too, and the tracks became impassable. For a month all traffic from Mombasa halted. Gangs of laborers were sent back from the Rift to rebuild washed-out embankments, reconstruct bridges, and realign tracks.

Ronald Preston faced a hard choice at the eastern wall where he and his work gangs were poised. The ground had become muddy and dangerously slippery. How could Whitehouse expect progress in such conditions? Preston was reluctant to advance

Workers lift rail into position in the Rift Valley. On the opposite page a temporary system hauls supply cars on the Rift Valley's eastern wall. (George Whitehouse is at the far left in the top photograph.)

down the slope into the Rift, but he knew that each day's delay put the railway further behind schedule. Slowly, and with less difficulty than he had anticipated, the crews mastered the first incline. But the second was steeper and more treacherous. The men, preparing the way with picks and shovels, could barely stand on the muddy hillside. Others risked their lives manhandling heavy crossties and rails down the slope. Many slipped and fell. The rails, each eight hundred pounds and thirty feet long, broke away, crushing bones and sweeping away everything in their paths.

The injured were carried to the top and removed for treatment, and Preston was forced to assign more men to the teams. These crews put crossties and tracks in place again, only to have them

slip down the incline. Finally Preston had to order men to sit on the rails and hold them down until they could be secured. The job was a hazardous one; no one was eager to squat on rails that could slide away and carry them down the cliffside.

The entire project had became a tedious, nerve-racking ordeal that tested Preston and his crews to their limits. But the workers, poorly paid and far from home, responded well to Preston's leadership; perhaps he aroused their pride. Gradually, tracks were laid down the slope to the valley floor, to Mile 373. The effort took months and was finally completed by May. It was a tribute to Preston's care and extraordinary skills that no one was killed during the dangerous descent. The temporary system he constructed carried freight and passengers until late 1901 when permanent steel bridges finally replaced it.

10

The Final Leg

While Preston and his crews were struggling to descend into the Rift, a new threat to the railway arose. The three million pounds allotted for the project were running out. To make matters worse, the Railway Committee's appeal for two million pounds more could not have come at a more difficult time. Britain was still supporting a war against the Boers, South Africans of Dutch origin; this effort aroused greater concern in Parliament than did a remote, expensive railway to Uganda.

When the matter came before the House of Commons, Henry Labouchere, the railway's outspoken opponent, was well prepared. He reminded his peers that he had predicted the project would consume five million pounds instead of the three that had been estimated. Labouchere went further. Hadn't he warned that the railway was an extravagant waste? Everyone knew that the evil slave trade was merely a pretext.

But Parliament had already committed so much to the railway—and it was more than half completed—that it would not reverse itself. Despite Labouchere's fiery criticism, the railway funding bill passed by a large majority.

News that the House of Commons had approved more money for the railway arrived just when Preston and his men reached the floor of the Rift Valley. Now the weather gradually improved, and the tracklayers advanced easily along the flat surface—past extinct volcanoes, past shallow Lake Naivasha, past mud-hut villages encircled with thorn fences, through Maasai land. The spear-bearing, red-painted *morans*, young Maasai warriors notorious for raiding the Kikuyu for cattle and sheep, stood by, curious, and watched while the workers laid tracks northward. The railhead advanced quickly.

Encouraged by this progress, Preston took time to hunt again. He tramped along the dry valley floor with his dog, Hero. Here and there were acacias and tall, leafless, cactuslike euphorbia trees. Preston went around clumps of whistling thornbush, so named for the sound of the wind passing through their hollowed pods, and saw hippopotamuses, masses of pelicans, flamingos, and other water birds around the valley's lakes.

The tracks turned and crossed the Rift to the western wall. This heavily forested slope, the Mau, rises some 2000 feet from the valley. To assure the proper grade, an approach ten miles long, consisting of some twenty-seven bridges, was needed to reach its summit.

The ascent began in October. The crews had to clear a path through thick brush and climb the wooded slope etched with sharp ravines. Heavy rains returned to slow their progress. Tracks sank in the soggy earth; there were frequent derailments. Though the men were eager to ascend to the summit, where they hoped to catch their first sight of Lake Victoria, they were still two miles short of their goal when the year ended.

It was early in 1901 when they finally reached the top, but Preston and his crews were disappointed that their struggle up the western wall was not rewarded by sight of the great lake. It was shrouded in mist. Meanwhile, the men shivered on the windswept 8700-foot summit. Each morning the workers, wrapped in their thin cotton blankets, lit fires to melt tea water, which had frozen overnight. Each morning, too, Ronald Preston, accompanied by his Indian helper and some aides, left camp to climb a cliff and try to catch a glimpse of Lake Victoria.

The search for a view of the lake was a disheartening ritual. Each day low clouds and mist hid it, and the party strained in vain. Then one morning, when they were beyond the summit and descending its western slope, the sky suddenly brightened. Preston, who had already begun to turn away toward camp, was pulled back by an aide, who pointed out a distant gleaming on the horizon. The scene became clearer. Beyond faraway hills and plains a patch of white glistened—Lake Victoria.

Descending the western slopes of the Mau, the railway encoun-

Epilogue

Two more years were needed after the ceremonial last spike to realign tracks and replace temporary structures. By then—1903—Port Florence had regained its original African name, Kisumu, and the cost of constructing the 582-mile-long line had reached 5½ million pounds sterling (almost 27 million United States dollars at that time). Thus, it exceeded even the estimate of the railway's most outspoken critic, Henry Labouchere.

What became of the people most involved in building the railroad (called Kenya Railways today)? In recognition of his extraordinary accomplishments, George Whitehouse was knighted by the newly crowned king of England, Edward VII. Sir George went on to manage a railroad in Argentina. In 1910, he retired to England and lived there until his death in 1938.

Ronald and Florence Preston had come to love East Africa; they made it their home and Preston became a successful businessman. He attended the railway's fiftieth anniversary celebration in Nairobi in 1945 and died in Kenya in 1952.

Lieutenant Colonel J. H. Patterson returned to East Africa from England in 1906 on a hunting expedition. He and his traveling companions took the train inland from Mombasa. It came to Tsavo in the middle of the night, but this unfortunate timing did not stop the colonel from waking his fellow travelers to show them the bridge he had built. Half asleep, the other passengers were not overly impressed. Patterson, the lion killer of Tsavo, lived in England until his death in 1947. The two famous man-eating lions he shot stand on permanent exhibit at Chicago's Field Museum of Natural History.

What became of the almost 32,000 Indians who labored to build the Uganda Railway? Their individual fates are unrecorded, but about two-thirds of them returned to India; of these, more than 6000 were sent home sick. Death, mainly from malaria, claimed some 2500 others. The balance, almost 7000, remained in East Africa; 2000 of them continued to work for the railway.

The Uganda Railway—the Lunatic Line to critics, "the most courageous railway in the world" to admirers—was an economic failure at first. The trade it was to carry was slow to start. Exports were limited by the small capacity of the few cargo ships on Lake Victoria, and the tsetse fly held up the development of Uganda. It caused such severe sleeping-sickness epidemics that by 1906, 200,000 Africans had died from it.

As George Whitehouse had foreseen, Kisumu harbor proved to be too shallow for larger steamers. This obstacle was eventually overcome; ships especially designed to operate in shallow waters were put into service, and later a hydroelectric dam was built, raising the level of the lake. But before these changes, not enough cotton, Uganda's chief export, was reaching Kisumu to support the railway or to repay the costs of constructing it. Other sources of income were sought. The Government decided to develop agriculture in the rich highlands near the Rift Valley. Crops raised on these fertile lands could be shipped by rail to Mombasa for export.

The British Foreign Office in London now exercised its power; it changed the boundary between British East Africa and Uganda. All of Uganda east of Lake Victoria became part of the East Africa Protectorate, also called Kenya, after its highest mountain. The British Government appropriated the fertile highlands and leased them cheaply to white settlers. Most came from the United Kingdom or South Africa, but some arrived from Australia, New Zealand, Canada, the United States, Germany, and Rumania.

The settlers were excited by the opportunity to lease inexpensive, fertile land to farm, land already served by a railway. Europeans, as white people in English-speaking Africa are called, believed they were bringing civilization to an almost uninhabited wilderness. The British Commissioner of the Protectorate from 1900 to 1904 wrote: "We are not destroying any old or interesting system, but simply introducing order into blank, uninteresting, brutal barbarism." The foreigners failed to understand that tribal societies and cultures, radically different from their own, already existed in the land.

The Europeans had little awareness or concern that the tracts of land they rented so cheaply had been taken from Africans. This issue became a source of bitter controversy between the British administration and the Kikuyu, Kenya's largest tribe. The British insisted that they never took for settlement any territory that was inhabited, only land that was empty. The Kikuyu disputed this claim, saying that they had left their farms and homes only temporarily when famine and disease swept the area from 1898 to 1900. As settlers occupied more and more of the fertile farmlands, the area became known as the white highlands. The Kikuyu felt like displaced persons in their own homeland. Many were forced onto crowded reserves. Others lived on portions of the settlers' lands as tenant farmers.

The white settlers prospered. Kenya, the unwanted stepchild Great Britain had acquired as a railway corridor from the sea to favored Uganda, was soon rivaling, even exceeding, its neighbor in value to the Empire. By World War I (1914 to 1918), Kenya's plantations were producing coffee, tobacco, sisal, and sugar; its farms raised vegetables, wheat, barley, maize, and fruit. These crops were shipped on the railway and sold profitably for export.

The plan to create cargo, traffic, and revenues for the railway was succeeding. The railway, in turn, played a major role in developing Kenya. Nairobi became a modern, commercial center, and the railway was its lifeline. In 1920, Great Britain took the prospering territory into its Empire; Kenya became a colony of the British Crown.

For many years, Kenyans tried to express their wish for equal rights to the colonial Government and sought the return of lands that formerly had been theirs. In 1946, a Kenyan who had gone to England fifteen years earlier to plead for Africans' rights returned to his home. His name was Jomo Kenyatta. Kenyatta, a Kikuyu, found prosperity in his country. The population had increased dramatically during his absence. About 30,000 Europeans and 5½ million Africans were now living in Kenya. Despite the great black majority, however, power and almost all the wealth were in the possession of whites. The lot of the average African still was wretched. They were made to carry identification cards and were limited largely to menial jobs—manual labor on the white settlers' farms, building roads and other public works—and to lowly clerical positions. The Indians fared better. They worked for the Government, for the railway, and for businesses. They became shop owners, accountants, and tradespeople.

Kenyatta had been a mere boy when the railway came to his

Diesel-hauled freight train, 1965. Longonot, an extinct volcano, rises in the distance.

land. His grandfather had told him the ancient prophecy about the great iron snake, and to Kenyatta the prediction that white-skinned strangers would come and take the land away had come true. Clearly the coming of the railway had brought about a changeover to white rule. Kenyatta traveled throughout Kenya, urging his people to unite behind him in a movement to recover their lands and rights.

A long, hard struggle for justice and eventually for independence began. An uprising took place and was crushed, but the freedom movement Kenyatta led did not collapse. On December 12, 1963, sixty-eight years and one day after George Whitehouse arrived in Mombasa, Kenya became an independent nation. Jomo Kenyatta became Kenya's first president and served until his death in 1978. He had lived more than eighty years, long enough to see a railway turn his land from a near wilderness into a developing country. The Fire Snake had changed the lives and destinies of his people and had altered the course of history in East Africa.

Bibliography

Brown, Jeremy, *Kenyatta.* New York: E. P. Dutton & Company, Inc., 1973.

Churchill, Winston, *My African Journey.* London: The Holland Press, 1962.

Cox, Richard, *Kenyatta's Country.* New York, Washington: Praeger Publishers, Inc., 1965.

Eliot, Sir Charles, *The East Africa Protectorate.* London: Edward Arnold, 1905.

Foran, W. Robert, *A Cuckoo in Kenya.* London: Hutchinson and Company, 1936.

Hardy, Ronald, *The Iron Snake.* New York: G. P. Putnam's Sons, 1965.

Hill, M. F., *Permanent Way.* Nairobi, Kenya: East African Railways and Harbours, 1950.

Huxley, Elspeth, *Red Strangers.* New York and London: Harper & Brothers, 1939.

———, *White Man's Country.* New York: Praeger Publishers, Inc., 1968.

Jackson, Sir Frederick, *Early Days in East Africa.* London: Dawsons of Pall Mall, 1969.

Kenyatta, Jomo, *Facing Mount Kenya.* New York: Vintage Books, 1965.

Marsh, Zoe and G. W. Kingsnorth, *A History of East Africa.* Cambridge: Cambridge University Press, 1972.

Miller, Charles, *The Lunatic Express.* New York: The Macmillan Company, 1971.

Moorehead, Alan, *The White Nile.* New York: Dell Publishing Company, 1960.

Patterson, J. H. *The Man-Eaters of Tsavo.* London: Macmillan and Company, Ltd., 1927.

Index

** indicates illustration*

About the Author

Francine Jacobs was born in New York City and has a B.A. in Education from Queens College. She has taught in the elementary grades and now concentrates on writing children's books. A travel enthusiast, Mrs. Jacobs finds that her trips often stimulate ideas for books. A trip to Kenya, East Africa, led her to write *Fire Snake*. Mrs. Jacobs and her husband live in Westchester County, New York. They have a son and a daughter.